W9-DIF-185

The Nuclear Era

The Nuclear Era

Its History; Its Implications

Carl G. Jacobsen
University of Miami,
Coral Gables

Oelgeschlager, Gunn & Hain, Publishers, Inc.
Cambridge, Massachusetts

Spokesman
Nottingham, U.K.

International Standard Book Numbers: 0-89946-158-1 (U.S. cloth)
0-85124-346-0 (U.K. cloth)
0-85124-347-9 (U.K. paper)

Library of Congress Catalog Card Number: 82-8077

First published in the United States in 1982 by Oelgeschlager, Gunn & Hain, Publishers, Inc.

First published in Great Britain in 1982 by Spokesman

Printed in West Germany

Library of Congress Cataloging in Publication Data

Jacobsen, C. G. (Carl G.)
The nuclear era, its history, its implications.

Includes bibliographical references and index.
1. World Politics—1945– 2. Atomic warfare. I. Title.
D843.J33 1982 909.82 82-8077
ISBN 0-89946-158-1 AACR2
0-89946-165-4 (pbk.)

Contents

Acknowledgments

This book grew out of a five-part University of the Air series for Canadian Television (CTV). I am grateful to CTV, for seeking me out, and to the series audience, for their gratifying response; it was the latter that sparked the enthusiasm and determination needed to plough on.

Special thanks go to my family, who had hoped to see more of me after my last book was completed. When the hope was thwarted, by this project, they did not complain, but rallied around with indispensable moral support.

My gratitude also to my sterling "secretary," Virginia Barton, whose unstinting aid was crucial to the manuscript's early completion; and to Ray Graves, one of our most promising Ph.D. candidates, who prepared the book's index.

Finally, a heartfelt thank-you to those colleagues and specialists, in the United States, Canada, and Britain, who saw drafts and chapter versions and urged me to go on. Their encouragement and their obvious belief that the project was important sustained me through periods of doubt and exhaustion.

Introduction

In 1980 West German Chancellor Helmut Schmidt issued the somber warning that the international situation was becoming reminiscent of 1914. The chauvinist stance adopted by both superpowers during the early 1980s, and the conflict-prone context of increased North–South alienation, resource scarcities and competition, and dangerously inadequate lines of communication spawned as volatile a mix of conflicting passion and perception as that conjured up by the jealous rivalries of late nineteenth-century powers. The calm of summer 1914, a summer of glorious sun, vacation, and sang-froid, was deceptive. The logic nurtured by ingrained fears and established postures was like that of an avalanche waiting for a pebble to set it off. The pebble was the shot that killed Archduke Ferdinand in Sarajevo. Rigid railway schedules and mobilization timetables turned post-Sarajevo preparedness initiatives into an inexorably escalatory cycle of deterrent steps, a cycle that drove all to a war none wanted; dictates of prudence acquired an automaticity and a reverberating dynamic that allowed none to get off the boat. The military doctrines and capabilities of the early 1980s contained their own escalatory logic. The danger that they might similarly one day defy policy-makers' ability to retain control was indeed hauntingly evocative of that sunny August of peace denied.

According to some estimates, as of the early 1980s the world already harbored perhaps 50 times the nuclear "yield-effect" needed to destroy known society, and the momentum toward proliferation and further "arms race" spiraling appeared unstoppable. Pacifists who recoiled in horror clearly had both history and moral absolutes on their side. Yet government strategists and military planners could call on equally compelling and equally uncompromising logic. Former British government Chief Science Advisor Lord Zuckerman's contention that just one older Poseidon-type submarine (a minuscule fraction of current arsenals) ought to suffice for deterrent purposes, since it could unleash as much destruction as did all the combatants of World War II, was persuasive. On the other hand there were those, like William Van Cleave of the University of Southern California, who could proceed from different premises to the conclusion that a further quantum leap in Western arsenals was essential. Others again, like the noted British historian E. P. Thompson, found compelling cause to argue that the time had come for European nuclear disarmament, and for Europe to distance itself from both superpowers.

Dialogues of the deaf proliferated. Ignorance of the premises underlying opposing viewpoints often precluded constructive debate. Even the professional communities most intimately involved appeared peculiarly atomized, alienated from each other. There was astoundingly little communion between America's academic nuclear scientists (many of whom had been central both to early bomb-development efforts and to the strategic deliberations of the Eisenhower, Kennedy, and Johnson administrations) and their colleagues within the military-political and military-economic establishment of the early 1980s. And the Soviet situation appeared analogous. American "strategists" as a group appeared alienated from Western "Sovietologists." Communication between the disciplines was minimal, if not nonexistent. Historical reference points relevant to the analyses of both were all too often ignored. "Peaceniks," arms controllers, weapon systems analysts, diplomats, and historians—all tended to construct their own protective cocoons, insulating themselves and their arguments from the outside world and from each other.

The dangers inherent in logical constructs derived from narrow, functional premises is illustrated by the late-1970s American debate about Soviet birth-rate statistics. After decades of impressive progress, Soviet figures of the early 1970s showed an increase in infant mortality. Analysts jumped to the conclusion that this increase reflected rising alcoholism and health system deterioration caused by the diversion of funds to the military. Yet bouts of rampant alcoholism have been endemic in Russian society throughout its history, and the latter contention rested more on supposition than fact. The negative trend of the 1970s clearly owed more to two rather different phenomena.[1] First, the birth-giving group that dominated the statistics of this decade were those who had spent their early years during the immediate postwar period of scarcity and deprivation; females who have been undernourished when young will invariably be burdened by higher infant mortality rates than their more fortunate sisters. Second, Soviet conscription periods were cut during the late 1960s, from three to two years, and the call-up age was lowered from 19 to 18; the 1970s saw younger decommissioned soldiers returning to marry and impregnate younger sweethearts, girls who would be 18, 19, and 20, rather than two or three years older as had previously been the norm. Statistically, a teenager has less of a chance of giving birth to a healthy baby than does a 21- or 22-year-old. The point is that those who analyzed the Soviet data of the early 1970s had conceived of no reason to delve into the nutritional standards of the 1940s, nor had they conceived of conscription laws as a possible relevant factor. Nevertheless, both were obviously of vital importance to the issue at hand.

Better awareness of history could have a crucial impact on strategic policy-making. The immense defense budget increases of the early Reagan administration in Washington were premised in part on extreme estimates of the Soviet defense burden and the consequent calculation that Moscow would not be able to compete, that the Soviet economy would be burst asunder by attempts to match America's raising of the ante. Western Sovietologists greeted that argument with skepticism, but Sovietologists of note were excluded from administration policy councils. More to the point, perhaps, was the fact that even the starkest estimates of the contemporary Soviet defense burden paled when placed in the historical context. The most extreme postulates of the percentage of Soviet GNP devoted to defense were lower by an order of magnitude than the norm during periods of Russia's past. The 1660s to 1680s saw about 50 percent of the Russian government's funding go to defense; the figure rose to approximately 80 percent during the years 1710–1715; from the 1730s to the 1860s the norm was 40 to 46 percent, and a similar figure was reached as late as 1911. In fact, the lowest defense budget years of the late nineteenth century (as well as many during the first decades of Soviet rule) still saw percentages that exceeded the highest CIA estimates of 1980–81. The Soviet citizenry of today might have gone too soft to allow a return to 1680 practices; yet one noted authority suggests that an image of heightened American threat might serve to justify (and make possible) reversion to the practices of a less distant past.[2]

This book will attempt to draw together the concerns, interests, and expertise of diverse communities, from "strategists" to "arms controllers" and disarmament advocates. The prejudices and perceptions of opposing functional and ideological actors will be treated with respect, if not always deference. While the book focuses on current trends, a relatively thorough historical background section, combined with specific analyses of the particular viewpoints of each of the above-described groups, should make it a unique supplement to existing library holdings.

The world of the 1980s has become too volatile; the head-in-the-sand and ivory tower self-isolation of essential actors has become too dangerous. The possibility that conflict scenarios will run out of control because of misunderstanding and misperception is real, and unconscionable.

The author's peculiar background as a historian, Soviet area specialist, and strategic (defense) analyst and consultant should ensure the breadth of vision required to transcend functional interests and bonds.

And if optimum expectations and hopes are not realized, a first step will at least have been taken. Even if the book proves but a partial success, even if it lifts just a minute corner of the veil that protects various groups from the scrutiny of others, even if it starts but a hesitant dialogue, it will have succeeded. If it encourages understanding, and compassion, it will serve its purpose.

NOTES

1. I am grateful to Harriet Fast Scott for bringing these considerations to my attention. The relevant graphs look roughly as follows:

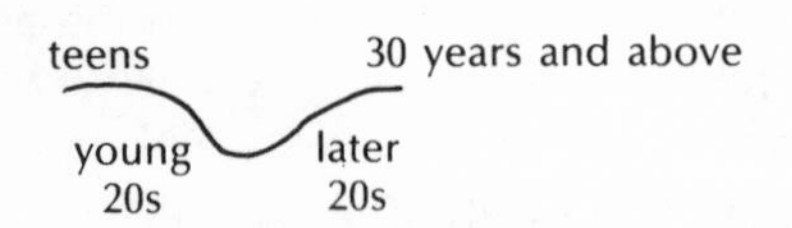

Infant mortality statistics as a function of mother's age.

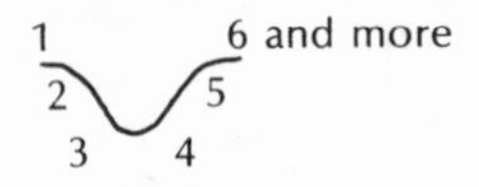

Infant mortality statistics per child born to the same woman.

The "infant mortality statistics per child born to the same woman" acquired increased significance as single-child family patterns began to dominate Slav demography. As concerns the 1980s, one should finally add that "the undernourished generation" would continue to affect statistics, even if only minimally: the ever fewer children still borne by this group would be affected both by the legacy of yesteryear and by the fact that the mothers would now be in the more hazardous age brackets for childbearing.

2. D. R. Jones, Director of the Russian Research Center of Nova Scotia (who also provided most of the data here presented), in discussion with this author, 20 October 1981. As noted by Jones, the figures and comments concern state budget percentages. The trend affecting GNP percentages is analogous, though firm figures for earlier years are elusive (in the absence of hard czarist government data, one would have to rely on somewhat uncertain extrapolations). Some of the figures provided are in fact obvious understatements. Typically, czarist figures exclude the very considerable cost of the gentry militia, a cost borne by the serf-peasantry rather than the state; nearly 50 percent of the agricultural production (and employment) may be counted a hidden defense expenditure. The 1911 figure, calculated by the Ministry of Finance, is unique in that it does include certain special funds, for strategic railways and shipbuilding. Yet all caveats notwithstanding, the fact remains that however specific figures are revised for purposes of comparison and compatibility, the conclusion *cum* suggestion in the text retains force.

Chapter 1

The Nuclear Era: Perception and Reality— A Century Apart?

The atomic devastation of Hiroshima and Nagasaki ushered in a new epoch of history.[1] The nuclear era had arrived. The atom would revolutionize calculations of world power. Its profound, cataclysmic impact dwarfed even that earlier watershed, the introduction of gunpowder. The years following Hiroshima saw the inexorable spreading of nuclear technology, nuclear weapons possession, and nuclear capability.

By 1980 the world harbored well over 50,000 nuclear warheads, many of a size that made the Hiroshima bomb look like a Chinese firecracker.[2] By 1980 at least six countries—the United States, the USSR, China, Britain, France, and India—possessed nuclear arsenals; two other countries, Israel and South Africa, were presumed (by Western intelligence agencies) to have at least some bombs; Pakistan was thought to possess the required technology; a number of states, including Brazil, Taiwan, and South Korea, were not far behind; altogether 40 nations had the knowledge and expertise to acquire nuclear arms within the next 10 years.[3]

The world military budget has shot past the $550 billion (US$) mark. Knowledge of nuclear phenomena has become so widely available in the open literature that students at major universities have been able to come up with workable bomb designs. That fact, and the evident laxity of security at certain bomb depots and, most particularly, in plutonium and nuclear "waste" storage and transportation methods makes it likely that nongovernmental groups, even terrorists, may become "nuclear-capable" in the future. (Astounding amounts of plutonium have been "lost" over the years—one might note that the CIA's estimate of current Israeli capabilities rests in part on the belief that one such shipment of plutonium ended up in Israel.)

Increasing North–South disparities, increased resource competition (especially for dwindling energy resources), and the failure of the old economic order to fulfill expectations serves at the same time to heighten the potential for conflict. The possible causes for friction, antagonism, and war have multiplied. The dangers are made even more acute by public apathy and widespread ignorance.

Today military budgets are still increasing, even more rapidly than in recent years—and this at a time when even a small decrease could have a startling effect. It has been calculated that just 5 percent of the world military budget could pay for the following programs:

vaccination against diseases that kill more than 5 million children annually

extending literacy by the end of this century to the 25 percent of the world's adult population that is unable to read or write and hence is cut off from most sources of knowledge

training health auxiliaries, barefoot doctors, and midwives (who can take care of 85 percent of a third world village's health needs) to service vast rural regions of the less developed world that have no access to professional medical services

eradicating the malnutrition that today sees more than 500 million people eating fewer calories than are needed to sustain ordinary physical activity, and that condemns 200 million preschool children to chronic hunger (one out of three children die from starvation before reaching the age of five)

providing supplementary feeding to 60 million malnourished pregnant and lactating women, that would dramatically decrease infant mortality rates

Even after paying for all this, there would still be enough money to establish 100 million new school places (250 million new school places are needed within the next half-decade just to keep third world enrollment at the 50-percent level), and to introduce hygienic water supply systems (water-borne diseases kill 25,000 people every day; such diseases are the most common cause of death among children under 5).[4]

All this for just 5 percent—yet military budgets keep going up, not down. Why? And how did we get here? That is what this endeavor hopes to explore.

Most people think of the immediate postwar world as one dominated by two superpowers, a "bipolar world." Yet Moscow's superpower status was hollow. It rested on the West's fear of the unknown, on the images conjured up by Zhukov's conquest of Berlin, . . Alexander I parading down the streets of Napoleon's Paris, cossacks, fables about "Asian hordes". . . . In economic terms, the Soviet Union was no superpower. Seventy percent of its industrial capacity lay waste. The Nazis' scorched-earth policy had taken its toll. Even with a slashing of armed forces personnel and forced-pace reindustrialization, it was not until 1949 that the USSR regained the industrial potential it had enjoyed in 1939. In the military sphere, Soviet inferiority was even more startling. The United States could obliterate Soviet society; but the USSR could not inflict damage on the US homeland.

Why should it have mattered? What drove the superpowers to a situation, a mere 35 years into the nuclear era, that saw each deploying more than 1000 times the explosive power of all the munitions fired in the Second World War? And in view of the obvious redundancy of such numbers, and the logical corollary that the larger portions of these arsenals appeared fated (if employed) to no more than stir the rubble, why then did even limited disarmament schemes remain beyond the realm of the politically possible? There is no simple answer to these questions. A number of complex, distinct, but interrelated phenomena and dynamics have helped shape events. Some would emphasize the arms race as traditionally perceived: an action-reaction spiral of weapon systems, each designed to offset or circumvent the effectiveness of the other's deployments, and each therefore viewed by the other as a threat to its relative power or potency and hence as cause and justification for a new round of systems improvement, ad infinitum. Others might emphasize rather the lure of evolving technologies (especially in societies whose prevailing ethics tend to associate "progress" with intrinsic good), and domestic factors like "bureaucratic politics" and interservice rivalries. (The American penchant for a triad of sea, air, and land-based strategic forces, and the tenet that each "leg" of the triad be independently able to devastate opposing coalitions, clearly owed something to the jealousies and lobbying power of the service branches.) Yet others would point to the fact that numbers of missiles acquired meaning at a time when quantity had to compensate for high failure rates, and that once relative numbers were sanctified as meaningful, it was inevitable that improved qualities led not to a corresponding lowering of numbers but rather to the promulgation of ever more number-devouring targeting doctrines and strategies (selective targeting; counterforce; withholding). Then there is the distorting logic of "lead-time," the time needed to research and develop new weapon systems, and its corollary: that budgetary allocations counter not what the opponent has but what the opponent might have 10 or more years down the line. The problem is exacerbated by the need for security planners to engage in "worst-case" prognoses, and hence to assume that the opponent might prove more capable than expected.[5] The awesome size and economic (and political!) clout of the respective "military-industrial complexes" ensures that each and every one of these exaggerating dynamics is allowed full and nearly unfettered play.

But there is still another phenomenon whose impact has been both pervasive and pernicious, a phenomenon which, though largely ignored, may have determined the course of events. It concerns the fact that while technology and means of destruction were catching up with the science fiction of the twentieth century, the mind-set of the protagonists remained mired in the prejudices of the nineteenth century. Thought and action patterns reflected the socialization processes of earlier eras. The media of all the major actors tended to be as myth-preserving and myth-propagating as the court bulletins of the Thirty Years' War. Each was inclined to caricature presumed opponents in the most prejudicial of words and images while cocooning themselves in self-righteous, chauvinist-cloaking self-portraits. The greatest fear of Karl von Clausewitz, the nineteenth-century "father" of modern strategic thought, focused on the destabilizing coincidence of a trend toward mass armies and mass destruction on the one hand and the combination of media manipulability, literacy, and democracy on the other; he dreaded the possibility that public passions and jingoistic impulses might usurp the role of the "strategist," and dictate the choice and course of conflict.[6] He would have abhorred the chancelleries of the 1980s, chancelleries which shunted aside the advice of nuclear scientists, preferring instead counsel from people with whom Czar Nicholas I and Teddy Roosevelt would have felt comfortable.

In analyzing both the misperceptions of others and the misperceptions of self, Russia—or rather its modern incarnation, the Union of Soviet Socialist Republics—provides a rich (though, as will be seen, by no means unique) lode of data. And in looking at modern Moscovy, it is as well to begin by looking at the Russian and Soviet obsession with security, partly because it bears so directly on balance-of-power calculations, but also because it remains inadequately understood in the West. There are two elements to this obsession. One combines living memories of the utter devastation and horror of the First World War, the Civil War, and the Great Patriotic War (which saw up to 90 percent of the Nazi war machine allocated to the Eastern Front) with cultural memories of earlier eras (245 external attacks between 1055 and 1462, including 200 assaults just between 1240 and 1426; of the 528 years from 1365 to 1893, Russia was at war for a total of 305).[7] As it is scarcely possible to find a family today that has not lost a member (or more) to war, so it has been in Russia for the last one thousand years.

The obsession based on ethnic and societal memories is further reinforced by the impact of Bolshevik ideology. It conditioned future Soviet leaders to expect that so-called capitalist nations would bend all efforts to destroy their Revolution, to snuff out the dogma that proclaimed capitalism to be the enemy of more egalitarian hopes and designs.[8] The Allied interventions after the Revolution, when US, British, French, Canadian, and Japanese troops landed on Russian soil, confirmed—indeed "proved"—the prejudice inculcated by ideology. Soviet leaders were also aware that final Allied withdrawal had reflected not goodwill, but rather resignation at the "White" armies' inability to secure sufficient popular support, together with a wave of mutinies in the French army, mass demonstrations in Britain, and a general war weariness that made prosecution of a war effort in Russia and Siberia a political impossibility. Later, depression hit the West. But there remained fear and conviction in Moscow that revived capitalism would try again. Nazi Germany was the first capitalist nation to revive, and true enough. . . . The war had seen other capitalist nations aligned with Moscow because they feared Berlin more. But the fall of the Third Reich would allow them to turn their energies against their ideological nemesis.

The postwar view from Moscow looked grim: US nuclear monopoly; US and Allied bases around the perimeter of the Soviet Union. The map as seen from Moscow, a Moscow steeped in paranoia, pictured encirclement, danger. The view of Eastern Europe also differed. It was not just the Bolsheviks who saw control over Eastern Europe as necessary. Non-Party Russians of conservative, moderate, and liberal bent agreed. East of the Tatra Mountains in Western Czechoslovakia the ground is flat all the way to Moscow, interrupted only by the Pripet Marshes. Over 20 million dead in World War II; over 20 million in World War I (and the ensuing Civil War); the ghosts of the Führer, the Kaiser, Napoleon, Charles XII, and others all intermingled. What came to be known as the Iron Curtain closely paralleled the line Catherine the Great had drawn on the map of Europe, as the line east of which Russia could not afford to be weak, could not afford not to dominate. The Soviet Union had to strive for the power to ward off all possible enemies, or else renounce both its ideological and its national aspirations. How could it pretend to leadership of an ideology that challenged the status quo if it resigned itself to perpetual inferiority? How could the Soviets attract allies if they could not help protect them? Britain and France could fall back on a powerful patron; they could rely on the US shield. Moscow saw itself as having no one to fall back on.

One might interject that Joseph Stalin's postwar probing for advantage in Iran, Turkey, Greece, and elsewhere contradicted assertions of weakness or reticence. But while Stalin was clearly ready to grab a falling plum or even pick one that was less than ripe (as long as he could do so with minimum risk), he shrank back from prospects of confrontation with his main protagonists. On this score, his posture was reminiscent of that of some of his czarist predecessors. The point should also be made that a degree of cold war tension suited Stalin's domestic purposes. During the late 1940s, as during the 1930s, the specter of foreign threat was a crucial ingredient of the policy of forced-pace industrialization (and its corollary, the deferring of consumer expectations); the appeal to patriotism crimped opposition rumblings. As an indication of attitudes concerning security, however, the fact of maneuverings on the periphery pales before the larger fact of basic caution, and the willingness (nay, eagerness!) to cut losses rather than face challenges.

There were actually two elements to Stalin's co-responsibility for cold war tensions and the resultant "arms race." The foreign probings were complemented by a determined effort to obfuscate the reality of Soviet military weakness and to promote an image of exaggerated strength—an image that fueled Western arms advocacies. The startling fact that Soviet leaders felt obliged to fabricate a myth so at variance with reality, in spite of its very real and negative side effects, testified to the depth of their insecurity and to their fear that the truth would invite Western attack. False strength projections, the flouting of potential "fifth-column" support within Western societies, and the espousal of nonconventional strategies (ranging from ideological subversion, through industrial espionage, to instances of sabotage and, *in extremis*, terrorism) were part and parcel of the manufacture of an image of deterrence, and hence a freedom to maneuver, that was as yet far out of proportion to their real strength. Success might make the catch-up task more arduous, but it would lessen attendant risks.

It has been suggested that Bolshevik ideology reinforced traditional nationalist inclinations on matters of security. The interplay between the two was in fact to become far more profound. A brief survey of the evolution of Bolshevism is essential to an understanding of this point.

One must first make a sharp distinction between Leninism and Marxism. A first-class intellectual, V. I. Lenin made a substantial contribution to Marxist literature (as in his writings on the nature of imperialism), but his drive to take and hold power also led to substantial deviations from the main body of Marxist thinking. It was Lenin who decreed the need for a single party of revolutionaries, a party that was to act as the revolutionary vanguard, and it was Lenin who insisted that this party must be organized on a hierarchical cell structure subject to the principle of democratic centralism. (There was to be no restriction on inner-party debates, but once decisions were made, minorities were obliged to adhere to and implement the directives of the majority.) Lenin agreed (!) with Rosa Luxemburg's warnings that his creation was open to authoritarian manipulation and therefore less than ideal. He argued that it was not to be a model for the future. It was time- and situation-dependent, designed initially to thwart penetration and disruption by the czarist secret police, and dictated later by the exigencies of war, and civil war. Lenin was to be accused by some of starting the Communist Party down the slippery road toward Stalinism when he sanctioned the crushing of the Kronstadt revolt (which he defined as counterrevolutionary and foreign-inspired) and when he lent his name to calls for discipline and loyalty that, per definition, narrowed the scope of permissible debate. There is no question that Lenin could be ruthless against "enemies of the revolution." Yet the situation was one of war and desperate struggle for survival. Wartime contingencies have always seen restrictions on civil liberties, even in the most "democratic" of nations.

It may be more important to remember that Lenin subsequently recoiled from the increasingly Stalin-dominant apparat of the Party (with more prominent Bolsheviks busied by "matters of state," Stalin had been assigned the job of managing a party swollen by idealists, opportunists, and cynics; his genius for backroom politics, intrigue, and patronage soon gave the innocuous administrator the rudiments of an organization).[9] Lenin's scathing attacks on Stalin's authoritarianism, suppression of dissent, bureaucracy, chauvinism, and general callousness came too late. Incapacitated by heart attacks, watched by doctors appointed by and beholden to Stalin, and subject to mail interception, the Father of the Revolution found himself in something akin to house arrest. Lenin was unable to secure Stalin's dismissal. But his attempt was testimony to the spirit that had given Kamenev and Zinoviev cabinet rank after they had committed the betrayal of releasing the Bolsheviks' October 1917 coup plans to the non-Party press. Lenin was always impatient with opposition, particularly during the desperate years of War Communism. Yet he had shown no vindictiveness toward colleagues who voted against, and sometimes outvoted, him after the Revolution. And his final writings hammered home the renewed conviction that the democratic component to the "democratic centralism" principle must be retained, or the idealism that remained integral to his vision would be lost.

Stalin came from a different mold. Ironically, his maneuvering for the succession was facilitated precisely by the assertion that the Party could not afford the disruption of open debate. Kamenev and Zinoviev, who like Lenin had deplored the authoritarian inclinations of Stalin's coterie, nevertheless saw Leon Trotsky's charisma as the greater threat, and therefore ignored Lenin's last appeal to join forces with the flamboyant "Permanent Revolution" advocate. Putting aside their misgivings about the apparat, they lent their names and prestige to a Stalin-orchestrated condemnation of Trotsky "factionalism." Underestimating the strength of the "law-and-order" appeal, and the depth of the yearning to relax after years of struggle, they helped cripple the power of their most natural ally. By the time they reverted to their original theme, they found themselves isolated, impaled on their own strictures against factionalism. The final act of the Greek tragedy saw the Stalin apparat secure the support of the dispirited remnants of radical sentiment in a purge of the remaining more conservative wings of the old Bolshevik leadership (including Bukharin, whom Lenin had described as "the darling of the Party"). Playing the rivalries and egos of bigger intellects like a master puppeteer, Stalin let his more famous rivals destroy each other. Revolutions devour their children.

By the mid-1930s all Lenin's old collaborators were off center stage, dead, abroad, or "retired."[10] Conformity became the byword both of the Party and of the society; the postrevolutionary explosion of artistic creativity withered with the advent of ever more narrow definitions of "socialist realism." The Revolution as originally defined was no more. The circumstances of Stalin's pursuit of power obliged him to assume the mantle of "Communist," much as the socialization processes of Spain obliged Francisco Franco to mouth the rhetoric of democracy. But Stalinism had as little to do with Karl Marx as Francoism had to do with Plato; to blame Marx for Stalin is indeed akin to blaming Jesus Christ for the Spanish Inquisition.

An early indication of what Stalin was, as opposed to what he was not, could be gleaned from his single "theoretical" contribution of substance—namely, his call for "Socialism in one country." This has often been seen merely as a pragmatic response to the failure of revolutionary aspirations in the rest of Europe, but it was more than that. It was a fundamental break with the internationalism of the early Bolsheviks. They had accepted Marx's argument that Communism in Russia was a contradiction in terms (the idea presupposed abundance, yet this reality mirrored only poverty and underdevelopment); to them the building of Communism in Russia would require aid from richer allies, allies with the means to put theory into practice. Their reach for power rested on the premise that revolution in Russia would spark the uprisings in Germany and elsewhere that in turn would nourish and sustain their endeavor. Stalin spoke to the increasing number of Party members who despaired of this scenario, or at least of its imminence, but at the same time he also co-opted the power and potency of Russian chauvinism. By declaring that the aim and purpose of the international movement must be to strengthen Soviet government, rather than vice versa, he in effect equated the purpose of the ideology with the purpose of Russian nationalism. And this was no tactical sidestep. Stalinist organization and Stalinist policy soon became ever more evocative of those of Ivan the Terrible (although it is relevant to note that this gentleman's Russian nickname was Grosnii, which connotes also a measure of awe and which is not wholly negative).

As forced-pace industrialization proceeded, Soviet society increasingly took on the form and character of an older Russia. With his castigation of Stalinist methods, the opening up of the camps and calls for a return to "the Leninist path" and "socialist legality," Nikita Khrushchev made a determined effort to revive some of the ideological idealism of yore. But his impetuous and impatient drive to reform anew, anew, and anew (without ever allowing the time to give any one particular scheme much of a chance), though marking a rather dramatic break with the immediate past, also unnerved the more conservative Party bureaucracy. There was no question of a return to Stalinist paranoia, but Khrushchev's successors did inch back toward the conservatism that had been so central to the success of the early Stalinist coalition.

Criticism of Stalin was focused more narrowly on "crimes" against the Party (in terms of relative loss, the Party had indeed been the single most devastated group in society), and offset by homage to his practical success in wrenching the nation through a process of telescoped industrialization. The emphasis on (socialist) legality was retained, and the trends toward greater egalitarianism and devolution of functional powers were allowed to proceed, but the operational bywords were caution and evolutionary gradualism.[11] The call for change was replaced by a call for stability—even where this threatened to melt into stagnation. The historian was reminded no longer of Ivan, but rather of Nicholas I. It was proper that one of the most daringly suggestive and controversial stage plays of the 1970s was to be a story of Pushkin, the nineteenth-century dissident and poet.

It is not that verbiage is unimportant, but that it really is extraordinarily difficult to differentiate the foreign policy effect of Moscow's late twentieth-century guise as bastion of "Communism" from that of the czars' self-image as guardians of Christianity. Both "manifest destiny" versions served to secure a limited degree of outside receptiveness to Moscow's blandishments, at least in some regions of the world. In areas dominated by others' colonialism, added benefit could furthermore be derived from the fact that Russia's particular brands of morality appeared less obviously tainted by hypocrisy. Moscow's current support for national liberation in areas of American political-economic dominance or sway echo the Russian navy's demonstrative visits to northern ports during the U.S. Civil War, at a time when British support for the South loomed as a distinct possibility.[12] Yet when ideological rhetoric jarred with national interest, the latter always ruled the roost (concern with Ottoman persecution of Christians, or the successor regimes' persecution of Marxists was not allowed to interfere with the larger designs of state). Given similar circumstances, there is every reason to believe that the czars would have reacted to events in Czechoslovakia (1968) and Afghanistan (1979) as did the Brezhnev regime. And there is every reason to suppose that they would have taken like advantage of opportunities such as presented themselves in Angola (1975) and Ethiopia (1978). The one difference concerning East European dissidence, and Polish dissidence in particular, may be that most of the czars would have resorted to force earlier and more harshly.[13]

There are those who argue that Bolshevik success in Moscow in fact retarded the nation's ability to pursue foreign interests. The argument rests partly on the historical record, that whatever the coloration of a power's chosen image of "destiny" (some variation of which is common to all), it always seemed to merge into particularly virulent chauvinism during the process of "industrial revolution" (final expansionist bursts of British and French colonialism are cases in point, as are the catch-up efforts of Kaiser Germany and Teddy Roosevelt's America); why should an industrially successful republican Russia have proved less prone to muscle-flexing temptations? A republican Russia would have emerged from World War I with a far more advantageous territorial position (the original post-czarist regime had been promised Constantinople in return for a prosecution of the war effort, and would surely also have picked up Tanganyika or some other of the German colonies that were parceled out among the victors (South-West Africa?!); furthermore, it would of course not have suffered the postwar losses of domestic territory that the Bolsheviks were forced to stomach). In addition, it would not have had to tailor action so as to at least appear consonant with an anti-imperialist ideology. Whatever the cause, there is little doubt that the policies of Soviet Russia have not evinced demonstrably greater assertiveness than did those of preindustrial Russia. Notwithstanding opportunism abroad, the pattern in both cases saw avoidance of distant commitments that might lead to direct clashes with the major rival(s), and in general an extraordinary willingness to cut losses in the face of opposition of substance (Moscow's reluctance to commit itself too firmly to Allende's Chile and the Sandinistas' Nicaragua and the years of caution that preceded the establishment of real commitments to Castro's Cuba exemplify the former; the quick acceptance of marching orders from former allies in China, Egypt, Sudan, and elsewhere is suggestive of the latter). In distant arenas, the pattern has in fact been that of the good poker player: cut losses and concentrate on the next deal, and the opportunities that it might bring.[14]

There is one advantage that modern Russia does derive from its pro forma ideological commitment. The emphasis on economic liberation and its corollary—that human rights aspirations are diversionary and counterproductive if not preceded by the establishment of job and social security guarantees—was clearly in accord with most of the world's thinking at the beginning of the penultimate decade of the twentieth century. But the perception of genuine ideological commitment of the kind that might have made allegiance to Moscow a matter of the heart had been whittled away by contradictory behavior in places like Egypt (where Russian chauvinism and boorishness had seemed little different from that previously exhibited by the British) and Zaire (where the promise of Western opposition sufficed to deter Soviet aid to "liberation" ventures). The image of assertively expansionist Soviet ideology contrasted with a reality more reminiscent of Machiavelli, a reality based on a cautious if shrewd nineteenth-century *realpolitik*. It was of course also a *realpolitik* that focused on the perpetuation of established power, a policy of conservative manipulation that throughout remained deeply suspicious of real change.

In China and the United States, nineteenth-century mind-sets appeared equally impervious to the theoretical dicta of officially espoused idealism-cum-ideology, though the discrepancy between theory and action was perhaps most blatant in China's case.[15] Contradictions affecting China's foreign policy are best known. Modern China's early aid programs both to selected Third World governments and to "liberation" groups elsewhere appeared uniquely sensitive to local custom, nondemanding, and altruistic, suggesting the primacy of ideological idealism over national self-interest (although it may have been indicative rather of sensitivity derived from China's own memories of colonial exploitation). Increased levels of involvement soon undermined this perception. In Vietnam the Vietminh, always ready to believe that old China's hegemonic impulses continued to lurk under the new veneer, soon found support for their cynicism. In Angola in 1975 it became obvious that national interest, then defined as anti-Sovietism, took precedence over any concept of social cause: China reacted to the pro-Moscow bias of the socialist-oriented People's Liberation Movement of Angola (MPLA) by spurning both the MPLA and the "moderate" alternative National Union for the Total Independence of Angola (UNITA) and instead throwing its support to the most reactionary of the so-called liberation movements, the CIA-sponsored National Liberation Front (FNL). China was well on the way to playing its part in that early-1980s travesty of respective ideological pretensions that saw China aid the "fascist" regime of General Pinochet in Chile, while Moscow nurtured ties with the junta in Argentina and Washington shored up the drug-running generals of Bolivia. The point here is of course not to judge, at least not per se, nor merely to observe that all powers sometimes resort to actions that belie their pretensions, but to emphasize once again the need to dig beneath the shibboleths of ideology and reexamine assumptions about the nature of contemporary decision-making and decision-makers.

In China's case, as in Russia's, such reexamination soon reveals the pervasive influence of history. In fact, the interrelationship between the two is illustrative. Their response to the border skirmishes of 1969 saw both referring to the single most potent common denominator of their histories (Jenghiz Khan), with Moscow depicting the Chinese as latter-day "Mongols" and Beijing scorning Russians as "Barbarians of the North" reincarnated; the suggested racism and cultural antipathy was in perfect accord with attitudes that have prevailed since their first official contacts in 1619. Supposed ideological affinity did little to dent a legacy of differing ethnocentric visions, conflicting territorial claims and aspirations, misunderstandings and antagonisms. The very concept of ideological affinity was from the beginning open to debate.

Mao's thinking had far more in common with that of the Taiping rebels of nineteenth-century China than with Karl Marx, whose works he may not have read. He was clearly impressed by Lenin's organizational precepts. And one might trace a debt to Trotsky in his later calls for permanent struggle (repeated "Cultural Revolutions") lest ideals be betrayed from within or without. He had seen the eclipse of Bolshevik leaders and the ensconcement in Moscow of a new mandarin class whose bureaucracy and chauvinism scarcely differed from that of the past, and he feared that his own designs might be similarly smothered. He was to smear opponents with the label of "Khrushchev's heir," thus rallying xenophobic reaction to Khrushchev's 1961 termination of aid, but the real object of his venom was Stalin. On the one hand Stalin had identified Soviet interest in China with the success of bourgeois nationalism (partly because he underestimated Maoist strength and potency, partly because he had no control over and distrusted Maoist ideology); his support had gone to Chiang Kai-shek rather than to Mao Zedong. In 1949, both needed the appearance of alliance to boost the image of potency in their deterrence postures, which otherwise appeared fragile (Stalin feared that a more fervently anti-Soviet America might use its strategic superiority, while Mao, who had been rebuffed by President Harry Truman in 1945, feared that cold war passions might tempt American support for a comeback attempt by Chiang from Taiwan). It was *force majeure*, not choice. After nine weeks of tough bargaining, Mao received military and economic aid, but he had to bind himself to a Soviet-type developmental model, pay with foodstuffs, raw materials, concessions in Manchuria and Port Arthur, and in general concede to Moscow special rights that made a mockery of the Karakhan manifesto of 1919.[16] Mao's vilification of Deng Xiaoping, whose policy prescriptions closely resembled those of the early Stalin regime, grew both out of the fact that Deng represented the centralizing antithesis to his "social-populist" vision, and out of the memory of Stalinist humiliations.

Deng Xiaoping's policies across the board, on Party organization and discipline, on economic planning and incentive procedures, as well as on the permitted scope and character of "individual rights," were indeed hauntingly parallel to those of the early Stalin years (and to those of Stalin's successors).[17] One is even struck by the parallel between Western perceptions of Deng and the initial Western reaction to Stalin's emergence; Stalin's ouster of Trotsky had been described as the victory of moderation over dangerous radicalism, and his encouragement of trade with the West had been hailed as the beginning of a new era.[18] But Deng's "Stalinism" also extended to the concept of "Socialism in one country," a concept which, as previously mentioned, defined the ideology as a servant of the nation-state, rather than vice versa. Deng's past relations with Moscow left little doubt that he viewed Sino–Soviet normalization as desirable for economic and other reasons, yet both sides clearly appreciated that the exorcising of ideological demons merely underscored the continuing relevance of historical and practical grievances, and that any negotiating process would be tortuous. In the meantime, Deng proceeded to open relations with Washington, in accordance with his preference for equidistant (and hence balancing) ties with all nation-state ("barbarian") rivals. That Sino–American relations thawed first was partly a function of US willingness to concede contentious issues like relations with

Taiwan, but it was primarily due to the realities of Beijing politics. The power of the Maoists had been dramatically eroded and they could no longer enforce their preferred policy of xenophobic implacability versus both superpowers, yet they retained enough residual strength to force Deng to defer (or at least proceed more cautiously with) his northern policy aims.

Notwithstanding functional similarities between Dengist preference and Soviet reality, Deng's underlying *Weltanschauung* owed its primary debt to the haughty traditions of China's old mandarin classes—as much a reflection of older dynastic tenets as was Brezhnev's. Stalinist analogies were largely coincidental to the more fundamental core of Dengism, much as analogies with Lenin or Trotsky, however apropos, nevertheless detract from truer appreciations of Maoist reality. In a sense Deng and Mao represented different strands of the same tradition. Mao was the inheritor and standard-bearer of Taiping and earlier rebel aspirations rather than those of established mandarins. But like previous beneficiaries of the Chinese belief that successful revolt against a corrupt or incompetent regime, *ipso facto*, conferred the Mandate of Heaven on the victor, Mao's exercise of subsequent prerogatives was to be increasingly divorced from the egalitarianism and utopianism of old. This indictment may be somewhat unfair, in view of the fact that Mao's final efforts focused precisely on the declared imperative of reversing his mandarins' inclinations to promote themselves as the New Establishment (it should be noted that although the attempt to return to the sociopolitical visions of the Revolution echoed Lenin's "last struggle," its roots are to be found in analogous attempts by the founders of earlier dynasties; the Cultural Revolution had clear antecedents in Chinese history). On the other hand, Mao's treatment of minorities and his reabsorption of Tibet, to name but two examples, reflected the pattern of ascendant dynasties of the past.[19] And there is no doubting the fact (and supreme irony) that Deng Xiaoping's success in the late 1970s was due in no small measure to the Maoists' attempts to force the pace of social change by dictating egalitarianism—a contradiction in terms that allowed their opponents to add the charge of hypocrisy to that of economically disruptive naiveté. Either way, the China of the 1980s was clearly more true to the China of earlier centuries than it was to the European roots of its ideological mantle.

The American dream, as utopian in its pristine state as the vision of either Lenin or Mao, was, as previously mentioned, equally vulnerable to the *realpolitik* dictates of great-power politics; and the dictates to which it was increasingly subordinated were again those born of nineteenth-century perceptions. The contradictions between ideals and reality were no less ironic than those that marked Soviet and Chinese policy postures. By the early 1980s Washington, like Moscow and Beijing, was perceived by much of the world to be pursuing actions that stood in sharp contrast to its proclaimed principles. The American ideal was the quintessential revolutionary and anti-imperalist creed. The product of the modern era's first national liberation war against colonial exploitation and magnet for European refugees from the tyrannies of state and prejudice, America as a globally involved superpower was in a sense a contradiction in terms (and one shared to a greater or lesser extent with all successful revolutionary regimes). But it was not merely a function of the revolutionary transformation from have-nots to haves, from challengers to power to custodians of a new status quo. One might argue that the roots were integral to the dilemma of all revolutionaries. The desperation that fuels revolt is a profoundly polarizing phenomenon, compelling participants to embrace visions that, because they must sustain through the darkest hours, are necessarily utopian, but which therefore also encourage expectations that no successor government can satisfy; revolutionary government cannot survive without compromise to its ideals.

In the American case the de facto betrayal of the myth was precisely what gave it its strength. The promise to refugees from afar was predicated on dead or dispersed Indians. The newest colonizers soothed their consciences by decreeing their predecessors heathen and unworthy of "civilized" privileges, but as with Lenin's justificatory "counterrevolutionary" epithet for Kronstadt sailors, the end result was to foster a tradition that struck at the very core of the motivating idealism. As the selective vision of the early Bolsheviks paved the way for the Stalinist perversion, so that of the white American settlers led to a glorification of frontier machismo that was to have a profound effect on attitudes to both domestic and foreign affairs. In the domestic arena it spawned an extreme form of early capitalism that sometimes rested more on the quality of hired guns than on business acumen. And it contributed to the conscience-soothing formula that linked failure with personal inadequacies rather than with shortcomings in theory or system. The glorification of the strong, which in domestic politics acknowledged no contradiction between equal rights and phenomena such as the developed world's least generous social welfare provisions (and poorest infant mortality tables) and big business' near-total control over news dissemination, was in foreign affairs to encourage a chauvinistic righteousness scarcely different from that exhibited by British empire-builders.

Ideals strong enough to propel revolutions may be subverted, but they are not easily forgotten. America's white ethnocentrism and its sometime fixation with cruder patterns of power manipulation, at home and abroad, found some of its sternest critics within the domestic social fabric. The larger idealism of President Woodrow Wilson's "Fourteen Points" after World War I remained vibrant in many sectors of the populace (as did the countering philosophical strand of isolationism). But Wilsonian egalitarianism and humanism, as expressed for example in the 1972 presidential campaign of George McGovern, tended to share the fate of Khrushchev's occasional calls for a return to earlier ideals. The Establishment at large dismissed them as naive, and hence as counterproductive to the purpose of effective policy-making. The perhaps simplistic idealism contained in UN Ambassador Andrew Young's policy statements toward Africa during the latter 1970s was not allowed to jeopardize America's ties with the Republic of South Africa;[20] similarly, sometime-expressed Soviet objections to the incarceration and worse of "Communists" in Arab nations were not allowed to affect state-to-state relations that were thought to be in the national interest.[21]

The Marshall Plan was perhaps the one US policy initiative during the post–World War II era that truly embodied the spirit of Wilsonian democracy. But the reason it secured legislative approval had as much to do with the fact that it was seen to entail a tremendous initial boost to US production, profits, and employment, and to promise an ultimately affluent market dominated by US technology and American financial and business institutions and practices. The Marshall Plan's mating of idealism and self-interest was unique, however, and behavioral patterns elsewhere left no doubt that self-interest would prevail in case of conflict.

A "massive and impressive" 1979 publication from Washington's Brookings Institution (the description is taken from *Foreign Affairs*) documented more than 200 incidents since 1945 in which the United States used the armed forces as an instrument of foreign policy-making.[22] And a number of these showed as blatant a disregard for the aspirations of the nations and populations in question as did Moscow's interventions in Eastern Europe. US aid was a decisive factor in the military overthrow of democratically elected governments in Guatemala (1954) and Chile (1973); the marine invasion of the Dominican Republic (1965) was likewise to thwart the verdict of democratic election; in Vietnam, US military forces were directed to combat a movement whose majority support was conceded by President Dwight Eisenhower and Secretary of State John Foster Dulles.[23] US military aid has of course also gone to a number of openly repressive regimes whose contempt for democratic ideals was absolute (and while some of these, like those of the Shah in Iran and Lon Nol in

Cambodia, were nevertheless toppled, others like those in Korea, Taiwan, Paraguay, Bolivia, Uruguay, and Chile continued to dominate Amnesty International and Church charts of the contemporary world's worst human rights violators).[24] Wilsonian idealism may not have been totally absent either from America's original assumption of hemispheric responsibility (the Monroe Doctrine) or from its decision to step into the post–World War II power vacuum caused by the demise of British and French empires and Russia's imperative focus on the tasks of reconstruction and consolidation. But Washington's jealous guarding of the fruits of historical fortune (and military-political predominance) was to be more reminiscent of perceptions and attitudes once associated with David Lloyd George and Georges Clemenceau.

The postures adopted by the two superpowers at the onset of the "second cold war" were starkly illustrative. The mileposts that marked the unraveling of détente aspirations during the 1970s, from the US Senate's refusal to ratify the 1972 trade agreement signed by President Richard Nixon and General Secretary Brezhnev to Moscow's intervention in Afghanistan in 1979, provided abundant evidence to those in each capital who sought proof of perfidy in the actions of the other. But this process of mutual alienation and disillusionment has been treated elsewhere.[25] The analysis here will concern itself only with the initially culminating phenomena, except to note that the very nature of these phenomena suggests that détente's fracturing may have been inevitable. The events of late 1980 and 1981 provided graphic example of both powers' readiness to set aside the ideals of ideology and propaganda whenever these jeopardized the dictates of more chauvinistic and traditional *Weltanschauungs*.

Moscow's response to the socioeconomic crisis and consequent slippage of government authority that unfolded in Poland throughout the fall of 1980 and into 1981 was not one-dimensional. But there was no doubting the bottom line: Soviet leaders made it quite clear that the so-called Brezhnev Doctrine of 1968 continued to apply; the "socialist community" had a right to concern itself with the affairs of a member state, and to intervene when and if ever it deemed such action to be necessary to protect the status quo. AFL–CIO and other American aid to Poland's independent trade union, Solidarity, and US broadcasts to Poland (through Voice of America, Radio Liberty and Radio Free Europe) were described as subversive and counterrevolutionary in intent, as inciting and incendiary. The rationale for intervention was thus prepared. If one stripped off the perfunctory ideological garb, however, one is left with the fact that Moscow was arrogating the right to define and enforce the limits to political deviation in Eastern Europe.[26] The point that needs to be made is that this determination and arrogance owed nothing to Marx; rather, it was a legacy of Russia's imperial past—Brezhnev's views of Moscovy's security requirements were no different from those of Alexander II, Nicholas I, and earlier czars.

The dynamics of Washington's posture toward the external world were hauntingly similar, in theory as in fact. UN Ambassador Jean Kirkpatrick developed the new Reagan administration's "doctrine" toward Latin America.[27] Their predecessor's occasional (and somewhat ambivalent) attempts to reemphasize the humanistic imperative of America's revolutionary heritage were dismissed as foolishly simplistic, indeed dangerous. Thus former President Carter's respect for Bolivia's democratic election of H. Siles Zuazo was scorned, as was his withdrawal of aid from the military junta that promptly intervened. (The aid termination was due to some of the officers' involvement in the cocaine trade, which was Bolivia's biggest export earner.) It was now declared that the "socialist" tinge to the election winner's aspirations should have sufficed to ensure US opposition to his investiture. It was furthermore suggested that such opposition should have been active; the United States must not only support military intervention against advocates of divergent philosophies, but also encourage and if necessary sponsor such intervention(s). The "Kirkpatrick Doctrine" was in fact the mirror image of the Brezhnev Doctrine, a cold and harsh assertion of imperial prerogative.

The doctrine was immediately applied to El Salvador. The opposition alliance in that country was declared to be a terrorist movement dependent on Soviet aid. Having thus "legitimized" action, the Reagan administration proceeded to provide arms and advisors to the military establishment. As Moscow apparently feared that unchecked Polish liberalization might infect other East European domains, so Washington postulated that opposition victory in El Salvador would encourage like aspirations in neighboring states. Indeed, this fear of the larger ramifications and consequences of inaction was clearly of greater concern than the supposed primary threat. Certainly the pro-forma pretexts met a chorus of outside skepticism: independent observers described American aid to Polish opposition as minimal, and noted that even Washington's own evidence of Soviet aid to the Salvadorian opposition showed this to have been indirect, and relatively minor.[28] Outside commentators saw both powers as artificially exaggerating the subversive role of the other in its own sphere in order to justify intervention. The essence of both crises was seen to lie in the bankruptcy of the domestic socioeconomic order, in the bankruptcy of the "client" regimes. The opposition movements in both cases were seen to be far more broadly based and far more representative than allowed by the respective powers. As concerned El Salvador and that context's peculiar viciousness, the leading news organs even of America's allies concurred with the Catholic Church's judgment that the country's security forces were responsible for the great majority of killings (at least 80 percent)—and that this pattern was precisely what had driven so many conservative and moderate leaders to embrace alliance with more extreme leftists.[29]

The fact that both Moscow and Washington were wont to follow the behavioral patterns of their imperial predecessors should occasion no surprise. But perceptions and behavior patterns rooted in the experiences and realities of earlier centuries may not be compatible with the fundamentally different parameters of the nuclear era. The socialized traditionalism of policy-makers and their practical and theoretical inability to divorce themselves from the legacy of past environments is in fact precisely what has allowed the nuclear genie such unconscionably unfettered freedom to mushroom out of control. As the famous doomsday clock on the masthead of the *Bulletin of the Atomic Scientists* crept inexorably closer to midnight, the ever more urgent need for consensus foundered on the contrary dynamic of ever more exclusive parochialism.

This dynamic was reflected in America's inability to take full advantage of its greatest practical advantage vis-à-vis the USSR. The fact that most of the world saw both superpowers' professed ideals as hypocritical, cynical, and self-serving invited a shift in focus to the *realpolitik* leverage inherent in America's greater wealth. There were those who argued that self-interest would ultimately compel revolutionaries to compromise with the single money market capable of providing substantive help, namely the Western (US-dominated) money market. But parochial lobbying within the US domestic polity restricted the availability of aid funds (to well below the UN "standard") and the parochialism of policy-making councils made the provision of funds so prejudicially dependent on compliance with US objectives as to court contrary rejection. On this score American policy was no different from that of Moscow and most other capitals, and their equally self-serving aid programs remained even more limited. But the fact was that the diffusion of economic power symbolized by OPEC was leading to a proliferation of donor and trade alternatives. Thus when Washington withdrew aid from the Sandinista government in Nicaragua in 1981, presumably calculating that paucity of aid from Moscow would force Managua to accept American policy dictates, Nicaragua proved able to secure major offsetting funding from Mexico (in particular), Western Europe, and Canada.[30] For the majority of nations there might indeed be no alternative to dependence on "capitalist" markets, but where these had been totally dominated by Washington during the early postwar decades, that was no longer so in the world of the eighties.

In the military arena the superpowers entered the 1980s more predominant than ever. With existing stockpiles of more than 1 million times the destructive power of the Hiroshima bomb, added to through continuous investments of well over $100 million per day, each had the yield-effect potential to destroy the other with but 1 or 2 percent of its arsenal.[31] Yet just as their hegemonic potential appeared most awesome in this sphere, their political and economic ability to implement possible hegemonic designs was becoming more and more cramped. And this was no longer merely a function of their mutual incompatibility. Rather, it was partly a measure of the degree to which the policies of each came to be viewed as parochial and exclusive; it was partly a function of the already-mentioned diffusion of economic power; and it was partly a consequence of the 1970s phenomenon that saw sophisticated weapons made so inexpensive and simple that their proliferation could no longer be controlled. The trend first manifested itself in 1973, when Egypt's still largely peasant army proved capable of utilizing surface-to-air (SAM) missiles, and it pointed directly to a future of far more lethally armed militias, guerrillas, and terrorists. The superpowers retained their ability to obliterate, but their scope for lower-level control options was being narrowed.

Ironically, one could argue that the danger of drift toward economic and military-political anarchy, the very danger that acted as propellant to nuclear proliferation trends, could be blamed on the absence of ideological conviction—and not, as generally presumed, on its presence. Through much of the less developed world, opposition to autocracy and/or outside dependence was welding what most had perceived as fundamentally disparate forces, namely, religion and Marxism. In Latin America the frequent juxtaposing of images of Jesus Christ and Che Guevara in rural homes was but one indicator of the radicalization of much of the junior clergy that occurred through the sixties and seventies; by 1980 significant sectors of the Church hierarchies (and erstwhile bastions of the Establishment) appeared to be treading the same path, especially in Brazil.[32] The trend echoed sometime assertions that Jesus Christ was the first Marxist, that the ultimate aspirations were identical, and that Marx's insistence on freedom from economic exploitation merely established the practical precondition to a fuller application of Christian principles.[33] It was this postulate, long accepted by Marxists such as Antonio Gramsci of Italy and Alexander Dubček of Czechoslovakia, that now increasingly found acceptance among Jesuits, Baptists, and members of other denominations. And the ideals of Marxism could also be fused with those of other religions, as witnessed by the case of the Mojahedin of Iran.

The startling fact about the above-described phenomenon is that the religious figures involved consciously disassociated themselves from US policy designs, and the Marxists equally consciously disassociated themselves from Moscow's. Surely both powers continued to attempt to infiltrate and manipulate the respective groupings. But their general inability and impotence to do so was a profound reflection of the cynicism that their past policies had generated. The ideals of their pro forma ideologies *could* be meshed; it is the chauvinism of their realities that cannot. And it is that chauvinism that has allowed the nuclear steamroller such free reign.

Postscript. This chapter was completed prior to the imposition of martial law in Poland, in December 1981. That act served as a sad but fitting exclamation mark to our analogy. General Jaruzelski's military regime appeared a carbon copy of the juntas of Central America (and as removed from the professed ideals of his outside patrons as were the latter from those of their guardian). The most determined core of Solidarity was rooted in the Marxist legacy of Rosa Luxemburg, a legacy nurtured in Gdansk, the shipyards and the mines; their ideals echoed those of Guatemala's and El Salvador's rebels. Poland's case was particularly poignant, because of its vibrant intellectual tradition; measured thus, in terms of what was lost, the most appropriate analogy might rather be sought in Pinochet's iron-fisted extinguishing of Chilean democracy. Still, the bottom line was that Warsaw appeared (at least temporarily) to have joined the ranks of "banana republics."

The most dangerous aspect of Poland's crisis may yet be the extent to which it has shown up, once again, the selectivity of espoused ideals. One remembers that many of those Americans who viewed "illegal" strikers in Poland as freedom fighters also applauded when "illegally" striking air traffic controllers in the United States lost their jobs, were prosecuted, and had their union "decertified." The abrogation of rights was far more extensive in Poland, of course, but then government-associated agencies in Guatemala were responsible for more deaths than was General Jaruzelski. Differences of detail may be less important than similarities of principle. Moscow cheers freedom fighters in Central America and gives them aid and succor, while orchestrating repression of the most elementary rights within her own sphere. Washington waxes equally indignant about events in Warsaw, yet continues to prop up the equally illegitimate regimes of guns and death squads to the South.[34]

The nuclear era cannot afford the czarist approach to Eastern

Europe and the world that Brezhnev seems to favor; nor can it afford the equally arrogant "Victorian" world views that Reagan appears to have embraced. Viceroy-type repression during the waning years of the twentieth century is not likely to be either as effective or as durable as in bygone times. Neither Poland, nor Chile, nor Guatemala (Czechoslovakia, Honduras . . .) promised to remain quiescent. "Mother-country" interventions promised to become more "necessary," yet also ever more likely to fail, or to produce ephemeral Pyrrhic victories at best. The probability that nineteenth-century mentalities would respond to frustration in a nineteenth-century manner was real, disturbing, and exceedingly dangerous.

Beliefs of convenience masquerading as beliefs of principle have always been dangerous; they have never been so lethal.

NOTES

1. The introduction is taken from this author's *The Nuclear Era: Its History; Its Implications*, a five-part University of the Air series for Canadian Television, aired December 1980–January 1981; published as opening chapter in Alva Myrdal and others, *Dynamics of European Nuclear Disarmament*, Spokesman, London, 1981.
2. The best sources on the realities of "the nuclear balance" are the annual *Yearbooks* and other publications of the Stockholm International Peace Research Institute (SIPRI), Ruth Leger Sivard's annual *World Military and Social Expenditures*, the monthly *Bulletin of the Atomic Scientists*, the publications of London's International Institute of Strategic Studies (IISS), and perhaps this author's work(?!).
3. William Epstein, "Ban on the Production of Fissionable Material for Weapons," *Scientific American*, July 1980.
4. Information taken from M. Thompson and E. Regehr, *A Time to Disarm*, United Nations' Association in Canada, Toronto, 1978. (The programs enumerated were in fact calculated to cost $17.5 billion, which represented 5 percent of the global arms budget for 1977; since subsequent global arms budgets have risen more quickly than inflation, one may presume that today's cost would in fact prove *lower* than that indicated).
5. C. G. Jacobsen, *Soviet Strategy–Soviet Foreign Policy*, 2nd ed., Robert MacLehose, the University Press, Glasgow (and Humanities Press, Atlantic Highlands, N.J.) 1974; and C. G. Jacobsen, *Soviet Strategic Initiatives: Challenge and Response*, 2nd ed., Praeger Publishers, New York, 1981.
6. Phil Williams, "Clausewitz: His Writings and Relevance," *RUSI*, London, 197–; and see Carl von Clausewitz, *On War*, Pelican, London, 1968 (with introduction by Anatol Rapoport), and M. Howard's "War as an Instrument of Policy," in H. Butterfield and M. Wight, eds., *Diplomatic Investigations*, Allen & Unwin, London, 1966.
7. Historian S. M. Soloviev, quoted in D. R. Jones's excellent "Russian Military Traditions and the Soviet Military Establishment," presented to USAF-sponsored conference on *The Soviet Union, What Lies Ahead*, Reston, Va., September 1980. To be published as part of conference proceedings.

8. The best study on the revolution is N. S. Sukhanov's *The Russian Revolution,* Oxford University Press, Oxford, 1965.
9. Moshe Lewin, *Lenin's Last Struggle*, Random House, New York, 1968.
10. Sukhanov, *The Russian Revolution*, and Lewin, *Lenin's Last Struggle*; see also Bertram D. Wolfe's *Three Who Made a Revolution*, Dial Press, New York, 1948, and D. H. Carr's *Socialism in One Country*, Macmillan, New York, 1958.
11. Jerry F. Hough and Merle Feinsod, *How the Soviet Union Is Governed,* Harvard University Press, Cambridge, Mass., 1980; see section on "The Policy Process."
12. J. W. Kipp, "Sergei Gorshkov and Naval Advocacy," in D. R. Jones, ed., *Soviet Armed Forces Review Annual*, Vol. 3, Academic International, Gulf Breeze, Fla., 1979. For other examples of czarist precedents as concerns foreign "adventurism," see Jones, "Russian Military Traditions." With regard to the continuity between the roles "guardian of Christianity" and "guardian of Communism," Jones suggests the reader compare M. I. Dragomirov's *Moments du Soldat,* Paris, 1889, p. 9; the Russian *Field Service Regulations* of 1912; and I. I. Yakubovsky, ed., *Boevoe Sotrudnichestvo bratskikh narodov i armii internatsionalist*, Voenizdat, 1975.
13. For first-class coverage of pre-1917 Russia, see N. V. Riasanovsky, *A History of Russia*, 2nd ed., Oxford University Press, Oxford, 1969.
14. D. R. Jones, conversation with author, at the University of New Brunswick's Centre for Conflict Studies, Workshop on "Low-Intensity Conflict and the Integrity of the Soviet Bloc," 24 March 1981.
15. For a fuller presentation of the data, see C. G. Jacobsen, *Sino-Soviet Relations Since Mao: the Chairman's Legacy*, Praeger Publishers, New York, 1981.
16. See J. W. Strong's fine survey, "Sino-Soviet Relations in Historical Perspective," in A. Bromke, ed., *Communist States at the Crossroads*, Praeger Publishers, New York, 1965.
17. Charles Bettelheim, "The Great Leap Backward," *Monthly Review*, July–August 1978.
18. *The (London) Times Atlas of World History* (Times Books Limited, London, 1978) put it this way: "Labelled an extremist, by 1925 Trotsky had been maneuvered out. . . . Stalin and Bukharin now dominated Soviet politics. Moderation, it seemed, had triumphed" (p. 258).
19. See J. K. Fairbank, E. O. Reischauer, and A. M. Craig, *A History of East Asian Civilization*, Vols. 1 & 2, Houghton Mifflin, Boston, 1965, for the best (Western) history of China.
20. See e.g., the *Washington Post* analysis in the *Manchester Guardian Weekly*, 29 March 1981.
21. An example of such objection may be found in *Pravda*, 31 July 1971.
22. B. M. Bleckman and S. S. Kaplan, *Force Without War: US Armed Forces as a Political Instrument*, The Brookings Institution, Washington, D.C., 1979. See also, e.g., R. I. Barnet's *Roots of War*, Atheneum, New York, 1972. Excellent background analysis is provided by J. C. Thomson, Jr., P. W. Stanley, and J. C. Perry, in *Sentimental Imperialists: The American Experience in East Asia,* Harper & Row, New York, 1981.
23. Neil Sheehan et al., *The Pentagon Papers*, Bantam, New York, 1971; see also G. C. Herring's *America's Longest War, The United States and Vietnam 1950–1975*, John Wiley & Sons, New York, 1979.
24. *Annual Report*, Amnesty International, London, December 1980.
25. For example, in this author's "The Changing American-Soviet Balance," *Current History*, October 1980. Note, e.g., George Kennan's commentary in *The Bulletin of the Atomic Scientists*, April 1980. See also Chapter 3.

26. See sources listed in *Studium,* April 1981 (news abstracts issued by the North American Study Center for Polish Affairs); also *The Manchester Guardian Weekly,* 12 April and 19 April 1981; and e.g., *New York Times,* 15 February and 6 April 1981.
27. See J. Kirkpatrick, "US Security and Latin America," *Commentary,* January 1981; see also her earlier "Dictatorships and Double Standards," *Commentary,* November 1979. Compare, e.g., *Latin America Weekly Report,* 25 July 1980 and 1 August 1980.
28. See, e.g., analysis in the *Manchester Guardian Weekly,* 15 March 1981.
29. See, for example, *Le Monde* and *Guardian* analyses in the *Manchester Guardian Weekly,* of 1 February, 1 and 22 March, and 12 and 19 April 1981; also, e.g., the (London) *Sunday Times,* 22 February 1981; the (London) *Financial Times,* 4 February 1981; *Stern* (Hamburg), 26 February 1981; and the editorial in *MacLean's* (Canada's national news magazine), 20 April 1981; and note analyses in the "Viewpoint" review section of the *Miami Herald,* 8 March 1981. The Inter-Church Committee on Human Rights in Latin America charges the military in El Salvador with responsibility for 85 percent of the killings, according to *Oxfam Canada El Salvador Appeal,* 1981. The *New York Times,* on 29 August 1981, covered France's and Mexico's decision to formally recognize the legitimacy of the Salvadorian opposition; see also *Miami Herald,* 28 August and 9 September 1981.
30. See *New York Times,* 10 May 1981, citing Mexican President J. López Portillo's attack on the "falsehood and sophistry in which the [American] campaign against Nicaragua is carried out in the name of democracy."
31. See Ruth Leger Sivard, *World Military and Social Expenditures 1980,* WMSE Publications, Leesburg, Va., 1980. See also *The Bulletin of the Atomic Scientists,* September 1979, pp. 32–43; January 1981, pp. 15–22, and March 1981, especially pp. 6–18.
32. See "Brazil Church's voice of dissent growing," in "Around the Americas," *Miami Herald,* 19 April 1981.
33. See, e.g., I. Meszaros, *Marx' Theory of Alienation,* 3rd ed., Merlin Press, London, 1972.
34. The analogy does of course extend also to the nature of officially sanctioned "elections" (*de facto* exclusion of genuine opposition parties; official monitoring of who votes and who does not, with the latter subject to sanction or penalty; and voting procedures that generally preclude private and unobserved marking of ballots).

Chapter 2

From American Nuclear Dominance to US–Soviet Strategic Equivalence

The enormous scale of Russia's war effort, culminating as it did in the storming of Berlin and final victory on a front so immense that it dwarfed Western theaters of combat, earned Moscow a world power (superpower) status that it had not enjoyed since the first half of the nineteenth century. But the cost was dire, in both human and economic terms. Seventy percent of its industrial infrastructure lay waste. Russia's armies were stripped of manpower; its labor and other resources were directed to a sometime frenetic reindustrialization effort. Not until 1949 did Russia regain its industrial strength of 1939.[1]

The United States, on the other hand, boasted the cataclysmic power of nuclear weaponry. American industrial capacity stood unscathed, and supreme. And access to British assets (the price extracted for wartime loans and aid) provided Washington new-found global reach and influence. There was only one true superpower. Washington's initial stock of actual bombs may have been minimal, but it had demonstrated the capability, and it possessed a unique fleet of strategic bombers. (The Soviet Air Force, developed as a tactical adjunct of the army, had nothing to compare.)

Moscow compensated through policies of deliberate deception. It did not reveal the extent of postwar demobilizations. To the contrary, by glorifying the massive sweep of its armies in 1944 and 1945, it encouraged its adversaries to believe that there remained forces that could be unleashed, yet farther afield if necessary. Like Mao later, Stalin belittled the importance of the atom while directing a top-priority effort to unlocking its secrets. The first longer-range bombers, acquired in the 1950s, would make flypasses over Red Square in front of assembled foreign dignitaries, double back, and fly past again to give the appearance of greater numbers. When the first few missiles were developed in the late 1950s Soviet leaders made fantastic claims as to their effectiveness (Khrushchev asserted they could shoot a fly out of the sky), long after it was realized that they were in fact so primitive and so faulty that few if any were likely to reach their targets.[2]

The Soviet policy of deception clearly encouraged Western overreaction. Lulled by Stalin's apparent nonchalance toward nuclear weapons, and by their own complacent assumption that Soviet possession remained but a distant prospect, American politicians were jarred by the Soviet explosion of 1949—and promptly initiated development programs designed to counter a now-exaggerated specter of Soviet capabilities.[3] The pattern was to be repeated again and again during the following years. Moscow's sleight of hand on the bomber question led to the first (and then utterly unrealistic) wave of American fear about Soviet "first-strike" capabilities, and a new policy of permanent alert for the Strategic Air Command. Khrushchev's empty boasts about missiles and his associated denigration of bombers fueled the Kennedy administration's strategic procurement programs of the early 1960s, programs that added more than a thousand missiles to America's already awesome arsenal of strategic bombers.

The so-called alliance with China in 1949 may be seen in a similar light. There was little affinity between Stalin and Mao. Stalin distrusted Mao's purposes and inclinations and had scant faith in his prospects; right up until 1949 he had preferred to channel the major portion of Soviet aid to the bourgeois nationalist government of Chiang Kai-shek. The pro forma ideological bond between Stalin and Mao was akin to that which had once existed between the German kaiser and the British king. Its importance paled when national and other interests diverged. The 1949 negotiations in Moscow were exceedingly tough, highly pragmatic, and at times acrimonious. The fact that they were finally concluded was due to both sides' perception that their respective security needs dictated at least the appearance of amity. The increasingly anti-Soviet bellicosity of American politics made for jitteryness in a Moscow mindful of continuing military weakness; in Beijing, meanwhile, the blanket and undiscriminating nature of

Washington's anti-Communism was seen to increase the danger and possible potency of a Chiang comeback attempt from Taiwan.[4]

The appearance of Sino–Soviet concord did nothing to dilute America's fear. As with Soviet force exaggerations, its immediate impact could be argued to have been counterproductive to Moscow's determination to acquire offsetting military potential—a determination dictated both by Moscow's profession of ideological leadership and (perhaps more importantly) by Slavophile readings of history. But these mirage components of Soviet strength did clearly serve to keep the West "off balance." The length to which Soviet leaders were prepared to go to secure that goal, and the costs that they shouldered in the process, testifies to the level of fear and unease that pervaded Soviet policy councils. As noted elsewhere, the insecurity was deeply rooted in Russian perceptions of history and stoked by ideological prejudices, prejudices "confirmed" both by Allied interventions during the Civil War and by Hitler's Barbarossa campaign.

The predisposition in favor of mistrust led inevitably to a cynical reading of the motives for America's late-war and early postwar policy initiatives (especially as concerned the economic management of occupied Germany).[5] And it influenced Moscow's ultimate opposition to the Marshall Plan. The Plan came to be seen, at best, as an attempt to generate a captive market tied to American technology and know-how (in effect, a "neo-colonialist" dependency), at worst as a vehicle for the creation and support of appendage armies.[6] Western liberals' abhorrence of Russia's imperial assertiveness found no sympathy in Moscow, where the westward move of the nation's border was seen as the return of a birthright, a birthright usurped by Western fiat after 1917, and where the imposition of hegemony in Eastern Europe was viewed as the answer to basic historically defined security requirements (as seen by Russian rulers going back to Catherine the Great).

The assertion that Moscow suffered from apprehension, insecurity, and perhaps paranoia appears amply justified by any reading of the confluence of military-industrial weakness and Slavophile and Bolshevik predilections, the more so in view of the insistent attempts to manufacture a larger-than-life image of deterrence. But it is still further reinforced by a review of other Soviet initiatives of the period. Having secured Eastern Europe, Stalin turned to pursue also the southern ambitions and probings of his Czarist predecessors. In the process he tried to wrest back control of the Russian-held parts of Armenia that the young Soviet government had been forced to cede to Turkey. But substantive American opposition to these ventures led to quick Soviet withdrawal. Yugoslav critic M. Djilas later recounted Stalin's frank appraisal to confidants: explaining Moscow's resignation in the face of Washington's aid to Athens in 1947, at the time of the Greek Civil War, he noted simply that paucity of Soviet military power ruled out other options.[7]

Other events worthy of note in this context include Stalin's late espousal of peaceful coexistence: confrontation was seen to have become too dangerous; the need to avert conflict was seen to dictate a more conciliatory posture.[8] This was followed, after Stalin's death, by L. P. Beria's suggestion that Moscow would permit German reunification if assured of that country's neutrality.[9] In 1955 Khrushchev sanctioned just such a *quid pro quo* in Austria.

The propagation of a political image of peace advocacy and reasonableness was of course part and parcel of the same deterrence posture that decreed the continuing inflating of military and military-political prowess. Under the circumstances there is every reason to concede a measure of sincerity to Moscow's political posture. Nevertheless, one must also note that it served to soothe, if only slightly, the Western arms procurement responses occasioned by military and alliance rhetoric.

Before we turn to a chronological account of the evolving "balance of power," one final thought deserves mention, namely, that the arms race impetus occasioned by this rhetoric may not have been its only cost. The rhetoric may well also have been primarily responsible, if only indirectly, for the Cuban missile crisis of 1962. Moscow's deployment of shorter-range missiles to Cuba has been viewed alternately as a move to secure Cuban independence (President Kennedy's assurances on this point allowed Moscow to withdraw with a semblance of "face"), or, more generally, as a cheap way of augmenting limited Soviet delivery capabilities in the face of rapidly expanding American means (the obvious failure of any such design has in turn been seen as a primary propellant of Moscow's more generous defense funding of the later 1960s and 1970s). Appreciation of Moscow's strategic inferiority at the time, a point to which we shall return, provides compelling logic to the argument that the second consideration was a significant causal factor. But the timing may not have been due solely to a dispassionate reading of Kennedy's dramatic procurement program, and the dilemma thus forced on Soviet security planners increasingly aware of their own limitations concerning both quality and quantity. The trigger may instead have been occasioned by a shift in American official attitudes, from apparent initial awe of Soviet missile accomplishments to considered public disdain for the products that they had spawned. Growing Soviet awareness of missile limitations and vulnerabilities was one thing. The apparent realization thereof by Washington was a different matter. To the extent that lingering fear of American intent persisted, the statements must have had an unnerving (and precipitating?) effect on Soviet leaders.

The point is that vulnerability, weakness, and fear of negative trends have precipitated far more conflicts than have assurances of might. Most of history's major wars were initiated not by nations assured of their power, but by nations who feared that inaction might erode what residual power they had, nations that felt driven to secure the advantage of the initial blow.[10] Unfortunately, this is another one of those lessons of history that tend all too often to be ignored by policy-makers and opinion-leaders alike.

Turning to the balance of power per se, one might start with the successful culmination, in 1949, of Moscow's crash program to develop a nuclear bomb. In and of itself, it made scant difference to Soviet security planners. Moscow still had no sure way of getting it to US soil. Through the 1960s and 1970s the irrelevance of bomb possession (ever easier to secure because of the increasing availability of the relevant scientific data), if not accompanied by meaningful delivery capabilities, provided the main deterrent to nuclear weapons proliferation. (Until the explosion of sophisticated arms sales, including fighter-bombers and cruise missiles, in the mid- and late 1970s, delivery capabilities had been far more difficult to secure.) But in 1949 such considerations were more than offset by the knowledge that then-existing intelligence means left a large margin of uncertainty in an opponent's calculations. Appreciating the political gain to be derived from an early explosion, Moscow could feel fairly confident that the outside world's ignorance of the woeful state of the Soviet backup potential would suffice to deter hostile action.

Reality, however, continued to haunt Soviet security planners. Even after the Soviet Union managed to acquire some limited bomber capacity, during the ensuing years, its ability to strike the US homeland remained questionable. Western analysts might have had to assume the worst, in the time-honored tradition of defense communities, but the view from Moscow must have conceded real doubt about Soviet bombers' ability to penetrate Western defense systems.

The ICBM, the intercontinental-range ballistic missile, was clearly the trump that was to release the USSR from the straitjacket of inferiority. Realization of its shortcomings and the consequent manipulation of a smokescreen to hide these from hostile eyes came later. There is every reason to presume that the initial Soviet reaction (as also the initial US reaction) to the first successful test flights was that their goal had been realized. Soviet leaders clearly felt that a revolutionary advance had been effected. Typically, they began to speak of the navy as a military concept of the past. But as suggested earlier, they were soon in for a rude awakening. By the turn of the decade it was realized that the navy, far from having become obsolete, was in fact crucial to the missile's future.

This was probably the most dangerous period of postwar history. Herman Kahn thought it was "an accident that an accident did not happen." The few missiles that Moscow was able to deploy (perhaps 30 by early 1960) were of such uncertain efficacy that one doubts whether the Soviet command could have felt confident that more than two or three, maybe five, would ever arrive anywhere near the designated targets. And that calculation rested on the uncertain premise that they might get off the ground. They were of course stored above ground and thus were highly vulnerable to enemy bomber strikes. The Soviets knew that in a crisis they would have to strike first if they were to have any chance of inflicting even limited punishment on the United States; Washington, on the other hand, if appraised of the true situation, would know that if American forces struck first there might be no damage to US cities. A dangerous "temptation syndrome" affected the calculations of both sides. We now know that Moscow chose, as a matter of policy, to err on the side of caution throughout these years. In fact, there was little choice. Command and control procedures were as rudimentary as the missiles. Permanent alert would inevitably have led to accidents, especially in view of the haphazard state of the art of available radars and radar analysis techniques (see below). Permanent alert was in any case impossible during the early years, since missile fuels were also in their infancy; they were prone to explode and had to be stored separately from the missiles. Still, the fact that Soviet missiles needed a week to 10 days to get off the ground (pushbutton readiness was indeed a myth) did not make the temptation syndrome any less real. It only increased Moscow's agony.

Deficiencies in intelligence gathering and radar data processing made matters even worse. US forces suffered a number of false alerts; flocks of Canadian geese were repeatedly "seen" by radar scanners as incoming Soviet bomber squadrons. And Soviet facilities were no better. Satellites, new radars, and better back-up procedures soon improved the situation. But the dawn of the missile age was particularly harrowing.

Faced with mushrooming US force potential, Moscow apparently decided that it could not or would not match the American mix of a dramatic numbers increase and simultaneous improvements in quality. This decision (taken about 1960), like the decade-earlier decision not to try to match American bomber capabilities, was surely in part a recognition of the fact that Soviet assembly-line techniques remained inferior to those of the United States. In part it clearly also reflected Khrushchev's priority concern with the raising of civilian living standards. There was no question of conceding American supremacy. But Khrushchev clung to the belief that nuclearization of the forces and modern rocketry could be fused into a leaner, less expensive, yet sufficient deterrent. It was a belief that was to contribute to his undoing, after the Cuban crisis and its denouement.

There were two elements to the Khrushchev regime's efforts. One focused, naturally, on ensuring the survivability of the limited forces at its disposal. Missiles were emplaced in protective silos, and aboard submarines. Improved fuel types were developed, allowing missiles to be "tanked up and ready to go." Mobile—and hence less easily targetable—missile systems were also built. Possible satellite deployment was toyed with (and later scrapped, as it was realized that the predictability of satellite trajectories afforded too many advantages to hostile intercept designs). Command and control facilities were greatly improved.

By the mid-1960s Moscow could be said to have developed a secure "second-strike force," a force that could with confidence be expected to survive an enemy strike and be available for a retaliatory "second strike." Only then could Moscow be said to have become a genuine superpower. The Soviet arsenal still paled beside that of the United States. But the fact that the Soviets were now indisputably able to incinerate much of America's urban population gave them a leverage that they had not previously enjoyed.

The "temptation syndrome" was no more. We had entered the age of mutual assured destruction (some prefer the acronym MAD). MAD—the guaranteed ability of either superpower to obliterate the other, even if the other attacks first—remains the essence of today's strategic balance. And it is likely to remain so for the foreseeable future.

The remainder of the 1960s and 1970s saw a determined Soviet drive to match also the wider panoply (and options implications) of US strategic power. The Strategic Arms Limitation Talks Treaty of 1972, SALT I, came to symbolize Soviet attainment of strategic parity. The treaty placed a ceiling on the number of missiles that each side was allowed, giving Moscow a sufficient edge to compensate for continuing American bomber superiority.

To explain why it took some seven years after the establishment of MAD before Washington proved willing thus to concede its existence as a basic and (at least in the short and medium term) lasting condition of US–Soviet relations, one must address the problems and prospects of the 1960s' most notable arms innovations. One was pioneered by the Soviet Union, as an integral part of its effort to secure the survivability of available missiles. This was the concept of ballistic missile defense (BMD).

Actually, the first tentative steps toward exploring the possibility of ballistic missile defense can be dated back to the very earliest days of ICBM testing. One is tempted to explain the alacrity with which Soviet scientists at least began to think of an "antidote" to the offensive missile as a function of their ideologically derived familiarity with the theory of dialectics (the theory that each "thesis" finds its contradicting "anti-thesis," and that the clash between the two forces the development of a "synthesis," which in turn becomes the new "thesis," now confronted by its own internal inconsistency, the new anti-thesis, ad infinitum). The Slavophiles' inculcated obsession with defense may also have provided its own rationale and impetus. But the more substantive developmental effort probably owed more to the realization of offensive system faults and vulnerabilities. One might suggest that as the original Soviet focus on the ICBM was due to inability to match American bomber efforts, the focus on BMD could be seen as an attempt to circumvent the implications of American missile procurement rates through the development of alternate technology. Attempts to short-circuit arms deficiencies through innovation was in fact a prominent feature of many facets of Soviet military planning. While Soviet technology remained generally inferior to that of the United States, Moscow again and again proved adept at unearthing previously neglected potentials (the pioneering of ship-to-ship cruise missiles, evolved from Germany's V-1 designs of World War II, stands as another example).

The first prototype Soviet BMD facility was in place by 1961. Three years later operational deployment began, around Moscow. Subsequent efforts centered on improving the Moscow complex. Original BMD technology was primitive. American data indicate that it was highly cost-inefficient, and of doubtful efficacy. Whether a BMD was calculated to save 10, 20, or, say, 50 million lives, the offensive addition that the other side would need to saturate the system and restore the previous status quo was bound to be cheaper. Soviet officials (and American advocates of BMD deployment) countered that no nation could forever hostage its survival to the presumed sanity of an opponent; a nation must pursue what indigenous means it could to ensure its own safety, even if initial costs proved disproportionate. And in fact their optimism that technological developments would favor the newer concept soon appeared to be vindicated. American data of the later 1960s showed that while complete protection remained infeasible on both technological and economic grounds, more modest schemes were rapidly becoming cost-effective.[11] The first American system was authorized.

But just as limited American defense deployment became a reality, the more substantive Soviet effort was stymied by an unforeseen roadblock. This roadblock was the result of America's developmental priority of the early 1960s, the perfection of offensive missile means. The first simple single-warhead missiles soon gave way to missiles carrying so-called penetration aids (PENAIDs) or decoys, designed to confuse enemy radar ability to identify the warhead and direct possible interception attempts. But these inexpensive decoys did not duplicate a warhead's composition, and their velocity and path through the atmosphere therefore diverged. Fears that enemy radar would be able to distinguish the real from the false led to the procurement of ever more expensive and sophisticated decoys, until the process culminated in the logical solution of employing multiple warheads—these would surely saturate interception capabilities. The solution, multiple re-entry vehicles (MRVs), was in effect a cluster of warheads that once released would fall free-fall. Then fear developed that the opponent might perfect longer-range defense intercept missiles that could intercept incoming booster rockets before the release of the warheads, an eventuality that conjured up return to square one.

The way out of the impasse was provided by MIRV (multiple independently-targeted re-entry vehicles) and its alter ego, MARV (maneuverable re-entry vehicles), missile concepts that allowed for far earlier warhead release and separation. A MIRV missile carries what is called a "bus" with a number of warheads that are released one at a time; after each release the bus realigns its trajectory and velocity in accordance with the particular coordinates of the next target. MARV differs in that its bus does not itself do the targeting, but remains instead on a steady course, while released warheads employ their own separate propulsion and guidance systems. The net effect is similar. That effect is demonstrated by the fact that early MIRV "footprints" (the total target area that could be hit by warheads from one single missile) could be represented as an oblong some 1,000 km long and 300 km wide. The merest difference of 1 meter per second (m/sec) in the release velocity of the warhead translated into a 1,600-meter difference on impact area. (For purposes of comparison: the original boost velocity would be about 5 km/sec; variations in individual warhead velocities would normally be on the scale of 10 to 20 m/sec).

The arms race spiral has sometimes been presented as a direct cause-and-effect, action-reaction syndrome. Early Soviet BMD efforts (which some blamed on the "excessive" size of the American arsenal) are said to have caused the American development of decoys, which is then seen to have spurred Soviet improvements in radar and intercept capabilities, improvements that in turn forced the United States to procure ever more sophisticated decoys, and MRVs, a turn of events that obligated Moscow to develop longer-range intercept capabilities, thus necessitating American funding of MIRVs, and so on.

Although superficially attractive and logical, however, such explanations do not suffice. Observers privy to American decision-making at the time inform us that Soviet capabilities rarely determined the outcome of policy deliberations. Decisions as to weapon characteristics and size (such as the decision to forgo higher missile booster numbers once the figure of 1054 had been reached, in favor of a concentration on MIRVs and other qualitative means of improvement) were often surprisingly arbitrary. Bureaucratic politics, interservice lobbying, the dynamics inherent in ongoing developmental programs, and perhaps a socialized compulsion to strive for (and, if possible, beyond) technological limits—all seem to have been more weighty on the scale than Soviet prowess, or lack of it.

The conflicting trends of technology left their mark in a shifting sands–type phenomenon that manifested itself during the early years of SALT negotiations. Preliminary discussions (1966–67) had seen an American demand that negotiations focus on limiting defense capabilities. Their argument professed to rest on the conviction that defense deployments would lead to a destabilizing arms race surge; far from seeing defense allocations merely as the redirecting of offensive funds into a less provocative field (a favored argument of BMD proponents), Washington insisted that real-life uncertainties would propel the other side toward a similar investment, and then compel both also to fund offsetting increases in the respective offensive arsenals. Pursuing the logic of this point, American officials went on to suggest that such an outcome would be doubly unstable, since it would inevitably induce fear that one or the other might calculate that its defense could absorb those of the opponent's missiles that would survive a first strike—in effect, a new variant of the temptation syndrome of yore. The psychology appeared sound. Nevertheless, the one-sided nature of the American stand was at the same time *ipso facto* detrimental to Soviet interests, since defense was the one area where they had invested the greatest effort, and where they consequently enjoyed a certain advantage.

Mirror-imaging Washington's selectiveness, however, Moscow blithely proceeded to profess interest only in restricting offensive capabilities. The Soviet argument was of course that there was no inevitability associated with the American specter, that its very premise could be voided through conscious decision-making, and that consequential restrictions on offense would dilute the fears of both sides. The argument was not devoid of logic. But its practical consequence was to discriminate against America's very significant lead in offensive weaponry.

Formal negotiations were put off for a while by the 1968 Soviet intervention in Czechoslovakia. When they finally were undertaken,

Richard Nixon was President and the positions had been reversed. Now Washington asked for offensive limits, Moscow for defensive.

Why the quick shift in attitudes? Well, as noted earlier, 1966–67 saw the Washington administration gradually being forced to give up its sole concentration on offensive capabilities. Domestic critics pointed to Soviet defense efforts, and to the previously mentioned fact that the administration's own evidence now showed the cost-exchange ratio (the technical term for calculating the relation between offensive and defensive costs) to be moving in favor of at least limited BMD systems. Designs to lower expected fatality rates by 10 to 20 percent began to look attractive. BMD advocates insisted that with the emerging one-to-one relationship between defensive and offensive outlays, "one dead Russian equals one live American," and that under these circumstances it was morally imperative for the US government to concentrate on the latter. They won the day. And in view of the USSR's continued reliance on single-warhead missiles, American defense protagonists were right. The limited systems now authorized for deployment in the United States held definite promise.

But on the other side of the coin, the more substantive Soviet defense system was being emasculated by emerging US multiple-warhead technologies (cost-exchange ratios favorable to defense aspirations presumed opponent reliance on single-warhead missiles). Moscow therefore put its defense deployments on hold. Existing facilities were not dismantled, but new procurement was stopped; residual funding was concentrated on future-oriented research. The greater immediate effort was shifted to offensive aspirations. Assembly lines were geared up for a dramatic jump in offensive missile numbers, and money began to be poured into the development of a Soviet MIRV capacity.

The Soviet predilection for defense was not completely superseded. In an attempt to make up for some of the void left by the dashing of immediate BMD expectations, Moscow ordered a sharp increase in civil defense efforts and training (in 1967 and 1968). Classes on protective and other measures to be taken in nuclear and chemical-biological environments were made obligatory, for both school-age children and older population groups. Paramilitary training of civilians was similarly extended and systematized (this particular program was intended in part to make up for a decreed cutback in national service training obligations, but it also had clear civil defense ramifications). Shelter construction was stepped up. But it was no panacea. The following decade saw a stream of reports in the Soviet press about plant managers who falsified training reports in order to avoid the production loss that compliance with official obligations entailed. By 1980 the majority of the population still did not have access to modern shelter protection. The practical difficulties of fully evacuating major cities appeared as daunting as ever (and no answer had been found to the argument that evacuation attempts might themselves provoke, incite, and precipitate enemy attack, or to the suggestion that a restrained antagonist would retain the option of employing up-wind airbursts to counter the effects of population scattering). Soviet achievements in the civil defense domain were not insubstantial, and they might suffice to make a major difference in a confrontation with lesser nuclear powers. But their ability to dent the sophisticated penetration capabilities and impact potentials of a superpower was severely limited.[12]

The change in SALT postures toward the end of the 1960s merely reflected American recognition of the dynamism of the Soviet offensive procurement program and Soviet recognition that the American defense effort held extraordinary promise as long as Moscow continued to rely on single-warhead missiles. Soviet procrastination during the following years was rooted in the realization that this constellation of affairs gave disproportionate advantage to American strength.

This was the main reason why SALT I could not be signed before 1972. It was only then that the evidence of emerging Soviet MIRV mastery became conclusive. Soviet penetration capabilities were now assured regardless of American defense investments. American defense technologies had been checkmated, as had those of the Soviet Union previously. Both sides were now forced to accept MAD as an inevitable adjunct to then-current defense technologies. "Parity," though perhaps distasteful to one or the other, had become a fact of life.

But if parity was accepted as a fact of life, however distasteful, it also implied mutual acknowledgment that the other would not precipitate ultimate conflict. The apparent recognition that neither side could afford major war undermined the rationales that had previously sustained both confrontation rhetoric and defense industry funding. Under these conditions, détente became a political imperative that could not be denied. That may also have been one of the reasons for its ultimate demise. Both the underlying condition for détente, and the momentum generated by its acceptance, struck at the core of the most powerful financial (and hence political) pressure group in the modern world. They undermined the future funding prospects of its largest single employer (paymaster for more than 50 percent of world scientists!), one whose immediate constituency reached into the millions—representing every profession from laboratory technician, welder, and draftsman to infantry soldiers and sanitation workers—and whose larger constituency counted yet more millions—from service station attendants, hamburger dispensers, and grocery clerks dependent on the proximity of one of its military or civilian components to a bevy of lawyers, accountants, and government lobbyists. All people are nervous when their job security is undermined. All are predisposed to favor developments that promise relief from mortgage payment worries and the like. Realism, not cynicism, demands awareness of the military-industrial complex's interest in questioning the soundness of détente's underlying assumptions and raison d'être. But the dissipating of détente was a process of later years, and it was to be a multifaceted process. 1972 was the year of détente.

There were two main parts to SALT I. One concerned offensive missiles. A five-year agreement limited the number of permitted ICBMs and SLBMs (submarine-launched ballistic missiles) to levels reached by midsummer 1972. It thus allowed the United States a total of 1054 ICBMs and 656 SLBMs (on 41 submarines), while the Soviet Union was permitted 1618 ICBMs and 740 SLBMs (on 56 nuclear-powered submarines). ICBMs deployed before 1964, and "older" SLBMs, could be replaced by new SLBM launchers; this would move respective limits up to 44 "modern" submarines for the United States, and 62 for the USSR. No restriction was placed on modernization per se. The parties were to be free to pursue qualitative improvements (though Moscow was prohibited from deploying more of the giant SS-9 missiles; the 309 that were in place or under construction were left untouched).

Senator Henry Jackson and certain other dissident members of the US Congress soon found cause to attack the apparent numerical advantage conceded to Moscow. Secretary of State Henry Kissinger focused the administration's response on the US lead in MIRV technology and deployment, and the contention that geared-up Soviet production capacities would, in the absence of an agreement, have allowed Moscow to establish even greater preponderance. Both points were misleading. The expected Soviet deployment of MIRVs ensured that the American advantage here would be temporary. As concerns the latter assertion, it was clear that Soviet production rates were tapering off. Like Washington before it, Moscow had decided that further force improvements were better sought in the qualitative arena. In fact, in this sense SALT I was militarily insignificant. It was the political import of its symbolism that was vital.

In military terms the criticism made scant sense. The SALT figures reflected the discrepancy between Moscow's preferred concentration on just two forms of deployment, land and sea (a "dyad"), and America's preference for a "triad," a three-way spread of land, sea, and air forces designed so that each of the three "legs" was separately able to meet the nation's strategic requirements. In a sense SALT numbers therefore juxtaposed 100 percent of Soviet force elements to two-thirds of those of the United States.

Washington's MIRV advantage translated into an ICBM warhead lead of 4300, to 2090 for the Soviet Union. Counting the US bomber advantage, 2000 weapons to 420 for the Soviet Union, one arrives at total warhead/bomb figures of 6300 American, 2510 Soviet. Because of the degree of initial advantage and the gradual and lengthy nature of MIRV conversion programs, the US lead in absolute numbers of warheads was to be retained throughout the 1970s and into the early 1980s (though spiraling totals had long since diluted the relative import of this reminder of America's pioneering role). When discussing the theme of deliverable warheads, one should finally note that the American lead was in fact even greater than it appeared. Geographical and technical constraints (Soviet submarines had to travel farther to firing locales, and longer-range SLBMs were still in their infancy) meant that only 40 percent of Soviet subs could be in firing position at any one time, compared to 60 percent of US vessels.

Moscow's only absolute advantage in 1972 lay in ICBM megatonnage. The larger Soviet missiles could deliver a total of some 11,400 megatons, whereas the US force could carry only 2400 megatons. This was another point that Western critics of accommodation were quick to seize upon. Once again a consideration of air capabilities made a certain difference: US long-range bombers were capable of delivering about 16,500 megatons, whereas Soviet bomber capacity was restricted to 3600 megatons.

The crucial point to remember, however, is the yield-effect relationship between accuracy and megatonnage (accuracy is given in terms of CEP, or "circular error probable," which measures the radius of the circle, centered on a designated target, within which 50 percent of a particular missile type can be expected to fall). An improvement in accuracy by a factor of two (say, a lowering of a missile's CEP from half a mile to a quarter of a mile) has the same effect as an eightfold increase in yield. Increasing the yield still further brings progressively smaller dividends in real effect. Because of the greater sophistication of US accuracy technologies, America could rest satisfied with smaller and cheaper missiles. It was because they could not secure the same accuracies that the Soviets were forced to resort to the uneconomical and cumbersome expedient of compensating with larger boosters and larger warheads. In this area too, Washington's advantage was to last through the 1970s and into the 1980s (although here also steady Soviet progress was to whittle away its impact).

The United States could draw comfort from residual differences in quality and sophistication. But the significance thereof paled beside the degree of redundancy already established by 1972. Existing numbers had evolved in part from determination to overinsurance against the high failure rates of early missiles, and in part from the need to compensate for even the remotest chance of significant opposition BMDs. With newer missile types and the specter of hostile defense capabilities receding, London's prestigious International Institute of Strategic Studies felt able to assert that the numbers concealed

> the most important fact of all: that while the United States and the Soviet Union are entirely unable to disarm each other by a first strike against strategic forces, each has within its armory a number and variety of both delivery vehicles and weapons capable of destroying any conceivable combination of second-strike targets within the other's territory.
>
> . . . Whatever detailed calculations may be constructed, neither superpower can consider itself to have any significant advantage over the other in terms of freedom to engage in nuclear war without incurring obliteration.[13]

The demise of established BMD concepts had of course been ratified/formalized by the second major component of SALT I, the agreement to limit defense systems. Though subject to review at five-year intervals, this treaty was to be of unlimited duration. It therefore signaled both parties' expectation that the "situation of quantitative and functional symmetry between the two sides" would be "permanent."[14]

The BMD accord (often called the ABM Treaty, after the anti-ballistic missile which constituted the heart of early BMD concepts—though alternative intercept modes such as laser and high-energy particle beams were to come to the fore in later years) authorized the signatories to deploy two ABM sites of up to 100 launchers each, one around the respective capitals, the other to protect an offensive missile complex. In theory the Soviet Union could thus expand the Moscow system, which had remained stagnant at the 64 launcher figure reached half a decade earlier and add a protective envelope around one ICBM site; the United States could complete its ongoing construction at the Grand Forks, North Dakota, ICBM base and proceed to deploy also around Washington. Due to the perceived inadequacy of current technologies and public protest (there were those who thought that the building of a Washington complex would merely guarantee its targeting by an offsetting number of additional offensive missiles), it soon became evident that America would not exercise the latter option. Moscow evinced far more interest, since protection around one of its Siberian ICBM installations might also serve the secondary (?) purpose of guarding against putative Chinese capabilities. But the argument that such a defense would merely serve to redirect emerging

Chinese potentials to any of a number of other targets of equal importance stayed Moscow's hand. Two years later, with increased pressure to reaffirm the logic and impetus of détente—during President Nixon's last visit to Moscow—the ABM option presented a ready sacrificial lamb. Authorized ABM deployment was restricted to one site for each country, Moscow and Grand Forks.

The shadow cast by uncertainty regarding China's future course and potentials was in fact the main reason for Moscow's refusal to countenance a possible disbandment of its existing defense system altogether. It might have been forced to concede MAD vis-à-vis the United States, as the inevitable consequence of the existing state and balance of the offensive/defensive technological equation; Moscow recognized that there was little that could be done to thwart the other superpower's sophisticated array of penetration capabilities. But the Soviets were not ready to concede similar vulnerability vis-à-vis third powers. The United States may have felt scant reason to fear other powers at this time. But Moscow was looking at a vociferously hostile China and NATO-aligned Britain and France. These and other prospective nuclear-armed states were not within reach of multiple-warhead technologies. The Moscow defense system might hold little promise against US means, but it retained significant promise against such penetration capabilities as could be mobilized by other challengers. The Moscow system provided a degree of relative immunity against non-US challenges. It made for a decisive differential in a potential showdown with a lesser power, the more so in view of the disproportionate national importance of Moscow's role as the apex of a centralized governing structure and as the hub of industry, transportation, and communication. The survival of the Moscow defense complex and the continuing and extensive Soviet testing of evolving defense technologies were predicated on Soviet leaders' determination to retain this edge.

As concerns the most vexing worry, Moscow also made sure to retain an added element of insurance. The ABM Treaty allowed existing test ranges to remain. The main Soviet testing area was at Sary Shagan in Central Asia, a location of some convenience in the context of projected Chinese flight trajectories. Moscow's significant investment into BMD prospects during subsequent years was no doubt fueled in part by an acknowledged distaste for MAD in all its guises, by at least distant hopes that an answer might yet be found even to American penetration certainties. But the primary determinant was clearly a function of Moscow's conviction that it stood a fair chance of being able to perpetuate at least its edge over other countries' evolving means.

One other argument may have influenced Moscow's posture. BMD protagonists in both the United States and the Soviet Union had long argued that even if their aspirations were to remain an impossible dream in the superpower context, limited systems ought nevertheless to be deployed as a hedge not only against third powers, but also against the dreaded possibilities of accidental launch. Analysts at America's Hudson Institute had been particularly prominent in warning that a future primitive nuclear power with just a few bombs aboard a submarine (or scattered on an unknown number of merchant vessels) might otherwise enjoy the bargaining leverage of a superpower; they also cautioned against smug assumptions about the ability of existing control and abort procedures to guarantee against unauthorized launch. Washington officialdom appeared less perturbed about the latter eventuality, and seemingly confident that the former could be averted through recourse to other policy options (Israel's 1981 bombing of an Iraqi nuclear reactor reminded observers that these need not be restricted to diplomatic and economic means of pressure). Soviet leaders may have been more receptive to the accident argument, however. The fact that Moscow played a larger national role than did Washington might be presumed to have led to a degree of added sensitivity. The socialization process of Russia's past was not conducive to complacency.

In conclusion, one might reaffirm, first, that SALT I was of minimal military importance. Neither side had bargained away any desired weapon system. In a real sense, the SALT agreement merely confirmed and ratified both sides' preferred options of deployment and procurement. The only military implications of consequence lay in the partners' commitment to facilitate the unimpeded functioning of each other's monitoring capabilities (satellites), in the designation of base booster rocket and "fractionalization" (multiple warheads per missile) numbers that were sufficiently verifiable and that could stand as adequate yardsticks against treaty violation allegations, and in the establishment of a joint committee to adjudicate possible disputes. The improvement of communications was of course in itself no small matter. The strictures against impeding satellite intelligence-gathering functions was calculated to entail that henceforth the force elements that could be hidden from the other's view would constitute no more than a few percent of available force components, a fraction so small as to be insignificant against the backdrop of the force levels now sanctified. The logical consequence of this state of affairs was to minimize the degree of exaggeration that both sides had previously been forced to build into their estimates of the opponent's capabilities. One impetus to inflated arms procurement rates had thus been restrained, if not checked.

SALT I did, however, have a more important function. As previously mentioned, it symbolized both sides' appreciation of the apparent permanence of the military impasse in which they found themselves, and their consequent determination to turn away from the cruder forms of confrontation politics. Together with the US–Soviet Trade Agreement signed by Nixon and Brezhnev at the same time, SALT I stood as the quintessential symbol of détente.

The ratification of Soviet "parity" deserves additional comment. There was no question of de facto Soviet parity in the strategic arena, notwithstanding residual American qualitative advantages. And there is no question that the outside world interpreted Nixon's signature to SALT I as American recognition of the new reality and of its likely permanence. Moscow's superpower status finally rested on solid military realities. Yet Moscow remained a regional superpower. It might be able to obliterate most of the world from its home bastion. But its ability to make its presence felt on the global arena was otherwise slight. The USSR was beginning to develop distant trade and investment interests of real consequence. Its worldwide fisheries and its krill harvesting of Southeast Africa, for example, were becoming vital to the satisfaction of the nation's protein requirements. But its conventional power projection means, the military muscle to protect and extend distant interests, remained minimal.

The further development of both was to be one of the hallmarks of the 1970s. But in 1972 the USSR's ability to bring decisive weight to bear on foreign problems was still largely restricted to adjacent regions. It still could not present a credible alliance alternative to African and other Third World nations who might have become disillusioned with Western ties, whether for political, economic, or other reasons. At the dawn of the 1970s the credibility of Soviet protection abilities in these regions was still to be proved. The constellation that provided Western interventionary designs with a protective envelope of military impunity was soon to be changed, but it had not changed yet.

Finally, one other comment must be made about SALT: both sides focused their bargaining efforts on limiting the advantages of the other. This should of course occasion no surprise. The same pattern has been prominent in all of history's tepid attempts toward "arms control," and it is indeed an inevitable consequence and outgrowth of the parties' primary allegiance to their own security prospects. But this also had another consequence. The Nixon administration used the "bargaining-chip" argument to secure congressional approval of a number of military programs whose prospects had appeared dubious (the Trident submarine program was one such effort, which had appeared doomed by a congressional consensus that it was financially excessive and militarily redundant). The argument suggested that a spurt of financial allocations was necessary to force Moscow to the negotiating table, at which point the programs in question would be bargained away for concessions of substance; the implied promise was that they would never proceed to the deployment stage. In the final analysis, however, not a single one of these purported bargaining chips was ever to be bargained away. A skeptic might justly murmur that the process of SALT negotiations had in fact acted, ironically, as a spur to arms procurement programs. Military (and military-industrial) acquiescence to the agreements did not come cheap. It bespoke a lobbying power and potency that should have given pause to arms controllers and détente advocates alike.

A comment on the Vietnam War may be appropriate, as a postscript of some relevance.[15] It was a war of many ironies. The man (Ho Chi Minh) who had rallied Vietnamese resistance to French and Japanese colonizers, who had modeled his independence declaration on that of the United States, with American officers at his side and expecting American recognition and aid, fell victim to the Truman regime's implacable view of his ideological debt. The unleashing of American forces rested on the belief that Ho and his Vietminh movement were integral parts of a world Communist conspiracy. Yet Ho's early ideological rift with Mao had been as fundamental as was the chauvinist legacy of Vietnam's 2000-year struggle against Chinese hegemony; the awesome scale of superpower assault (however "controlled") forced Ho to accept a measure of Chinese aid and dependence, but the precipitous cutting of these ties once success was finally assured testified to the depth of underlying memories and distrust. But the more important irony, for our purposes, lay in the devastating impact that America's war against Vietnam was to have on American power and American prestige. The decade-long drain of between one-third and one-fourth of Washington's military budget (depending on whether Vietnam expenditures are included or excluded from the base) greatly eased Moscow's military catch-up efforts. The most enduring legacy of the war was thus to hasten the erosion of American superiority by facilitating Moscow's attainment of substantive parity. Had it not been for the Vietnam War, the symbolism of SALT I might have been delayed a decade or more.

The devastating political impact of America's war prosecution efforts, and in particular of the use of napalm and anticivilian (terror) bombardment techniques, also poisoned American claims to moral superiority, again serving Moscow's purposes. New generations in Europe retained their parents' antipathy to the designs of Moscovy, but they did not retain their parents' faith in the automatic sanctity of Potomac policy initiatives. Moscow's moral posture had gained nothing in absolute terms, but it had gained immeasurably in relative terms. And the gain was to be permanent. A near-decade later, when its Afghan venture drew a deluge of condemnation, Moscow could still draw comfort from demonstrators' equating of the evils of Soviet and American imperialism. A "plague on both your houses" attitude had become pervasive in many regions, regions that before had known but one bête noire.

NOTES

1. For a good general survey, see A. Ulam, *Soviet Foreign Policy 1917–1967, Expansion and Co-existence,* Secker and Warburg, London, 1968. For a more detailed history tracing the postwar development of Soviet military capabilities and Soviet strategic concepts, to 1972, see C. G. Jacobsen, *Soviet Strategy–Soviet Foreign Policy,* 1st ed., Robert MacLehose, the University Press, Glasgow, 1972; 2nd ed., 1974 (American agents: Humanities Press, Atlantic Highlands, N.J.). A. G. Mileykovsky, *Mezhdunarodnie Otnosheniia Posle Vtoroi Mirovoi Voiniy* (International Relations After the Second World War), Vols. 1–3, Izdat. Polit. Lit., Moscow, 1962, is a fine exposition of Soviet views and perspectives.
2. See, for example, Les Aspin, "Debate Over US Strategic Forecasts: A Mixed Record," *Strategic Review,* Summer 1980; also see M. Nacht's "Response to Wohlstetter," *Foreign Policy,* No. 18, 1975.
3. It is interesting to note that scientists working on the American nuclear bomb project were less sanguine than the politicians; the scientists did expect Moscow to be able to reach its goal within four years or so—that is, by about 1949.
4. Note C. G. Jacobsen, *Sino-Soviet Relations Since Mao: The Chairman's Legacy,* Praeger Publishers, New York, 1981, Introduction and Chapter 2.
5. Note M. Gabriel, *The German Peace Settlement and the Berlin Crisis,* Paine-Whitman, New York, 1960; and G. Kolko, *The Politics of War: the World and United States Foreign Policy 1943–45,* Random House, New York, 1960.
6. Mileykovsky, *Mezhdunarodnie Otnosheniia.*
7. M. Djilas, *Conversations with Stalin,* Pelican, London, 1969, p. 141.
8. J. D. Stalin, *Pravda,* 2 August, 3 October, and 26 December 1952; and see M. Shulman, *Stalin's Foreign Policy Re-appraised,* Harvard University Press, Cambridge, Mass., 1963, Chapters 10 and 11.
9. Tibor Meray, *Thirteen Days That Shook the Kremlin,* Praeger Publishers, New York, 1959, p. 28.
10. R. N. Lebow, "Soviet Incentives for Brinkmanship," *The Bulletin of the Atomic Scientists,* May 1981, pp. 14–21.
11. See, e.g., the U. S. Secretary of Defense's annual "Posture Statements" to Congress, January 1967 and January 1968, and D. G. Brennan's "The Case for Missile Defense," *Foreign Affairs,* April 1969, pp. 433–449.
12. There appeared to be little illusion on this score. A popular Russian joke goes like this: "What will you do when the warning comes?" "Wrap myself in a sheet and walk slowly to the cemetery." "Why slowly?" "We don't want to cause a panic, do we?"
13. *The Military Balance 1972–73,* IISS, London, 1972, p. 86. The above figures were taken from this publication's special appendix on "SALT and the Strategic Balance."
14. Ibid.
15. See J. C. Thomson, Jr., P. W. Stanley, and J. C. Perry, *Sentimental Imperialists: The American Experience in East Asia,* Harper & Row, New York, 1981; also Frances Fitzgerald's highly praised *Fire in the Lake,* Vintage Books, New York, 1972.

Chapter 3

Moscow's Emergence as a Global Power

SALT I and the US–Soviet Trade Agreement of 1972 appeared to cement the arrival and permanence of détente. Both superpowers professed to believe that offsetting strategic capabilities were a fact of life and that the balance of technologies and productive capacities mocked concepts of superiority; real superiority was said to be illusory, and unattainable. Yet both elements of this fundamental premise were soon to be subject to attack and erosion. As this process accelerated during the ensuing years, détente was gradually but inexorably emasculated of both content and raison d'être. With the demise of détente came the removal of its constraining effect on arms procurement and competition.

The withering of détente can be blamed on the confluence of a number of different factors and trends. Residual ideological antipathy, fear, and distrust played a role, as did the more general fact that different historical and geographic perspectives often drove the two sides to discrepant and sometimes discordant views of world events. The institutional dynamism of the juggernaut "military-industrial complexes" spawned by the preceding decades and the personal interests and security concerns of their managers and dependents served a distinct and possibly crucial role, sustaining and reviving biases of tradition and circumstance through unremitting focus on the requirements of their industry and profession—requirements that centered on conspiratorial evaluations of the opponent's means, intentions, and prospects. These generic obstacles to a further flowering of détente (indeed to the very existence of détente) encouraged military procurement and thought patterns that deliberately or coincidentally dramatized designated problem areas. The doubts, real or feigned, were in other words themselves largely responsible for developments subsequently seen to have sustained and vindicated their bias and impact. Some of the military and military-political trends in question can be ascribed to Soviet design, others to American. We shall address the former first.

Moscow realized that the peculiarly limited nature of SALT I's pro forma provisions gave a distorted (though politically opportune) image of relative strength, and that more comprehensive balance sheets testified to continuing American advantage in the strategic arena. One might well argue that Soviet leaders in fact saw America's signature to the agreement as a license to proceed with their obvious determination to flesh out a more genuinely offsetting global force structure, as indicating American resignation that the "threshold of significance" had been passed and that full realization of Moscow's long-term ambition could now no longer be thwarted. Moscow proceeded with established developmental programs. And it may be pertinent to acknowledge that subsequent US administrations, Republican as well as Democratic, were to concede that Moscow's policy choices did not contravene the carefully formulated terms of the 1972 treaty.

One of the main pillars of Moscow's military developmental program of the 1970s lay in its dogged pursuit of the kind of strategic potential that might provide the same flexibility and policy option range as the American arsenal provided Washington.[1] The program had three foci: The first lay in the introduction of MIRV, followed by the gradual phasing out of most single-warhead missile types. As noted elsewhere, the end of the 1970s and the early 1980s saw a continuing American warhead lead (due to the head start enjoyed by the American MIRV conversion effort), but by then the relative significance of the difference was submerged under the impact of the now far larger numbers on both sides. The second focus of Soviet aspiration lay in the research, testing, and ultimate deployment of more sophisticated targeting technologies. By the end of the decade the accuracy potential of deployed Soviet missiles remained far inferior to that of the US force. But ongoing tests bore witness to a successful developmental effort; the accuracy (CEP) of Soviet missiles deployed during the 1980s would clearly equal or at least approximate American standards.

The proliferation of Soviet warhead numbers and the steady improvement in Soviet missile accuracies were complemented by a move toward what might best be termed a partial "withholding" strategy. The strategy called for certain force elements to be withheld from initial exchanges, so as to be available rather as leverage for inter-war bargaining and war-termination purposes. It was a strategy of luxury that was predicated on the existence of surplus means. The days of limited capabilities, limited numbers, and poor accuracies had dictated concentration on city-targeting options; only thus could the force potential (and its deterrence effect) be maximized. Larger numbers, greater certainty as to missile effectiveness, and improved accuracy subsequently invited targeting also of nonurban industrial targets, military bases, and the like. Emerging realization of the phenomenon and effect of "fratricide" (the first detonation, whether on target or not, is likely itself to cause the premature explosion or destruction of accompanying and follow-up warheads) and its corollary—that the past practice of compensating for possible failures by assigning additional warheads to each target had been illusory—served both to counter residual inclinations toward redundancy targeting and, in the process, to free yet more warheads for other purposes. As the number of available warheads outstripped burgeoning lists of credible strategic targeting alternatives, the impetus for further numbers proliferation was restricted to three arguments. One rested on extreme specters of the degree of relative crippling (and the political-military consequences thereof) that might result if the opponent was to unleash the full force of its present and feared future capabilities in surprise attack. To the extent that this argument retained any credibility at all, it did of course serve to encourage emphasis on the warhead potential of the least vulnerable force elements, at sea or elsewhere. The second point related to aspirations for withholding and general reserve capacities. The third argument highlighted the possible use of strategic missiles for tactical purposes, to compensate for perceived deficiencies in one's conventional force structure. American enamorment of this idea was manifest already by the early 1970s; it gave new purpose to emerging strategic potentials, and it answered doubts occasioned by the transition to a (smaller) volunteer force structure.

Moscow concentrated first on satisfying the second concern, though its chosen course also served, coincidentally or not, to soothe the fears of those who continued to be perturbed about hostile first-strike scenarios. The choice of emphasis was perhaps obvious, since Moscow did not yet possess the accuracy technology demanded by the third concept, and since the conscript nature of Soviet armies provided a larger pool of manpower trained for conventional exigencies. Anyway, Moscow's preference led to a new stress on the navy.

Some background comment may be relevant. The original buildup of Soviet naval capability (the navy had previously languished as little more than a coastal defense auxiliary to the army) was clearly spurred primarily by the need to counter the strategic threat of US carriers bringing nuclear-capable fighter-bombers within range of Soviet targets. The strategic defensive requirement was given added impetus by the American development of Polaris-type submarines. The subsequent procurement of Soviet SLBM capabilities reinforced the need for expanded anti-submarine warfare (ASW) resources, this time to protect indigenous means. This third developmental determinant might also be categorized as belonging to the realm of strategic defense, if one believed in current deterrence theories and their intrinsic sanctifying of the idea that second-strike force potentials must of necessity remain inviolable, or one would face a volatile Pandora's box of dangerously destabilizing dynamics (mutual assured destruction). Sophistry aside, however, the point is that the logic of each of the above-described tasks compelled a progressive extension of Soviet surface naval reach and potency.

The longer range being built into American carrier-borne aircraft and into American submarine-launched missiles inexorably drew Soviet engagement requirements ever further away from home waters. So did the fact that Soviet SLBMs initially had limited range capabilities, and therefore had to be fired from close to enemy shores. Actually, this restriction, and the accompanying need to secure transit through ocean passages dominated by NATO's potent ASW capabilities, was soon recognized as an intolerable constraint. The immediate response was to explore and protect Arctic transit routes that circumvented chokepoints like the "GIUK (Greenland–Iceland–UK) gap." The longer-term response lay in the pioneering development of intercontinental-range SLBMs. The year 1972 saw the deployment of the first Delta-class submarines capable of hitting US targets from the security of home waters. This heralded a new era. By the end of the decade remaining Yankee-class vessels, the erstwhile backbone of the Soviet SNB (submarine nuclear ballistic) fleet, were being refitted with ship-to-ship cruise missiles.

As conversion to a Delta (and successor) SBN fleet proceeded apace, it soon became clear that Moscow was directing an extraordinary effort to securing the sanctity and inviolability of first the Barents Sea and later also the Sea of Okhotsk, in the Far East. The scope of the effort went well beyond what might have been required had Moscow wished merely to safeguard the vessels' ability to survive enemy first-strike attempts, so as to be available for second-strike purposes. The intent was clearly to secure vessel immunity for a considerably longer term than required by narrow interpretations of second-strike doctrine. The protective envelopes established ranged from an exhaustive array of underwater devices (which in the North were to extend well into Arctic waters), to new carrier capabilities (Moscow's first vertical takeoff and landing—VTOL—carrier was assigned to the Northern theater, and the second to Vladivostok, gateway to the Okhotsk) and significantly increased naval shore-based air power (including the potent long-range Backfires and an array of cruise missile–armed fighter-bombers able to operate from unconventional landing strips—and thus not tied to vulnerable airfields).

The two other main foci of Soviet military procurement and thought during the 1970s concerned the improvement of Moscow's conventional and "theater-nuclear" capabilities in Europe (and the Far East), and the development of more significant distant force projection means. The former saw rapid qualitative change. Previous NATO-provided numbers had flattered Moscow. NATO tank figures, for example, habitually included reserves in the Warsaw Pact tally but not in the Western accounting (although the United States, for example, kept, and continues to keep, more than 5000 tanks in reserve in Germany alone); nor did NATO balance later tank numbers against related helicopter and antitank capabilities. Similarly, NATO tended to omit reference to the fact that Moscow's large fleet of aircraft was composed of shorter-range, limited-capability planes. As the 1970s progressed, however, the gap between fiction and reality narrowed. The end of the decade saw increased numbers of more sophisticated longer-range and multipurpose planes, capable of deep interdiction and all-weather operation.

As concerns ground and ground-related forces, Moscow's European posture retained its emphasis on tanks, notwithstanding the lethality of NATO's mushrooming force of antitank missiles. Since the Soviets had researched, developed, and tested the very missiles whose success during the 1973 Mid-East War sparked the subsequent surge in NATO antitank missile procurements, they could hardly be deemed unaware of their potential. The answer to the apparent anomaly of continued Soviet tank emphasis lay in Moscow's expectation that large-scale European war would be nuclear, an environment far less propitious to infantry-based antitank designs; Soviet tanks and armored personnel carriers were designed to operate under nuclear and chemical-biological conditions, not just as assault spearheads but also as insulated modes of transport. With regard to conventional scenarios, Moscow chose to end-run the problem by developing a fleet of helicopter gunships that could sweep a path for accompanying tank columns.

By 1980 the Soviet force of helicopter gunships still had not equaled that of NATO and the United States (it was of course the United States that first developed the concept and use of helicopter gunships, in Vietnam). On the other hand, Moscow was taking the lead in the development of larger and more heavily armed versions. And the Soviets were not restricting their ambition to antipersonnel concepts. Helicopter-borne antisurface guns and missiles were complemented by new types of air-to-air missiles. Moscow appeared cognizant of the helicopter both as an anti-helicopter weapon and, perhaps even more importantly, as a platform from which to engage the larger spectrum of enemy jet capabilities. (The helicopter's unique ability to hide and dart among tree clearings and other terrain features, combined with the fact that ground clutter gave it radar and missile homing advantages vis-à-vis more lofty foes, provided this ungainly bird with an edge that was rarely fully appreciated.)

The Soviet's enthusiasm for the helicopter extended still further, however. They supplemented their multirole helicopter gunships with a rapidly expanding fleet of helicopter transports. And they instituted training programs to prepare regular motor-rifle troops to take advantage of the new conveyance (8 to 10 hours of special instruction was said to be sufficient for the average infantryman). The Soviets did not try to emulate the American development of a distinct and self-sufficient air cavalry. Their leaner paratroop divisions were designed rather as shock troops with minimal sustaining capacity, and as behind-the-lines saboteurs and commandos; sustaining follow-up was to be the task of helicopter-carried, leapfrogging motor-rifle formations. The Soviet preference was equally evident in the naval sphere. Soviet "Marine" forces were limited to shock and specialized tasks, while regular army troops were trained (also) to employ naval—and even merchant ship—means of transport; as in the case of air-mobile operations, it was the army that was to provide the sustaining backbone to more ambitious sea-borne landings. The Soviet soldier was becoming far more mobile and versatile. Still, the process, though dynamic, remained evolutionary in nature. Conceptual differences notwithstanding, it must be noted that in terms of hardware America's long-established lead in helicopter transport capabilities carried over into the 1980s, much shrunken perhaps, but no less real in absolute terms.

The final major object of Soviet developmental efforts during the 1970s was, as mentioned, the procurement of distant air and naval capabilities. The emergence of the former was heralded by Moscow's ability to launch and support a rather protracted airlift of supplies to Egypt during the 1970 War of Attrition. Limited long-range air support potential was demonstrated during the Angolan conflict of

1975. By 1978, during the Ogaden War between Ethiopia and Somalia, Moscow was finally able to demonstrate that it now possessed distant airlift capabilities of major consequence.[2]

The emergence of naval power projection means followed a similar pattern. The 1960s had witnessed the westward move of Soviet naval exercise and deployment patterns (from home waters, through the North and Norwegian seas and into the eastern Atlantic), the establishment of a regular presence in the Mediterranean, and the first forays into the Indian Ocean. The first truly global Soviet naval exercise, Okean I, was held in 1970. The next 10 years saw the gradual, steady emergence of new vessel types with ever more impressive range and armaments characteristics. It was in the early 1970s that US naval design bureaus first began to identify cetain Soviet hull and other design features as worthy of emulation. The quality, variety, and sheer number of surface-to-surface and surface-to-air missile batteries aboard Soviet vessels were also salutary. During the 1975 Angolan conflict, a Soviet "West African Squadron" was able to play a significant military-political role. By the end of the 1970s Moscow's navy was able to engage in simultaneous operations of military-political significance in a variety of ocean theaters, off China's coast, in the Red Sea and northwestern reaches of the Indian Ocean, and (to a lesser extent) in the Caribbean.

The development of global military reach was accompanied by the emergence of a more assertive strand of Soviet strategic thought. Interventionary war had previously been described as a capitalist phenomenon, a function of capitalism's innate need for captive markets. The early 1970s, however, brought acknowledgment that interventions might also be fueled by other factors. Concomitant with the emergence of this definitional flexibility came a whole new branch of Soviet strategic literature. One of the best-known Soviet strategic authors of the day, V. M. Kulish, led the way, in 1972. Having "secularized" the concept of intervention, he proceeded:

> Military support must be furnished to those nations fighting for their freedom. . . . The Soviet Union may require mobile and well trained and well equipped armed forces. . . . The actual situation may require the Soviet Union to carry out measures aimed at restraining the aggressive acts of imperialism. . . . expanding the scale of Soviet military presence and military assistance furnished by other socialist states are being viewed today as a very important factor in international relations.
>
> . . . . In some situations the very knowledge of a Soviet military presence in an area in which a conflict situation is developing may serve to restrain the imperialists and local reaction. . . .[3]

Naval Commander-in Chief, Fleet Admiral S. Gorshkov, pursued the theme. As summarized by a Western analyst:

> Gorshkov differentiates between two tasks in peacetime naval diplomacy—one directly involving the security of the USSR itself and its maritime frontiers, and the other involving Soviet state interests in the seas and in the Third World. . . . The second task doesn't directly involve USSR territory or Communist interests, but state interests, state economic interests in the seas—merchant marine, fishing fleet and especially the mineral resources of the sea water and sea bed—and Soviet state economic, political and military interests in non-Communist countries of the third world. Gorshkov speaks of protecting state interests as an "especially important" task.[4]

Gorshkov's assertion that the navy must be an active (and not merely passive) instrument of state policy in the global arena complemented the views of Defense Minister A. Grechko:

> The external function of the Soviet state and its Armed Forces and of the other socialist countries and their armies has now been enriched with new content. . . . At the present stage the historic function of the Armed Forces is not restricted merely to their function in defending our Motherland and the other socialist countries. In its foreign policy activity the Soviet state actively purposefully opposes the export of counter-revolution and the policy of oppression, supports the national liberation struggle, and resolutely resists imperialistic aggression in whatever distant region of our planet it may appear.[5]

Secretary General Leonid Brezhnev weighed in:

> Our Party is rendering and will render support to peoples who are fighting for their freedom—we are acting as our revolutionary conscience and our communist convictions permit us.[6]

Moscow frankly stated its view that détente entailed no "obligation to guarantee the status quo in the world or to stop the processes of class and national liberation struggle."[7] To the contrary: "Détente does not mean—and will never mean—a freezing of the social and political status quo in the world, or a halt to the anti-imperialist struggle . . . and against foreign intervention and repression."[8] The principle of coexistence "has nothing in common with class peace between exploiters and exploited, the colonialists and victims of colonial oppression, or between the oppressors and the oppressed."[9] Contrary Western concepts "only testify to a false understanding of the meaning of détente, which never implied and cannot imply a free hand to aggressors."[10]

US Secretary of State Henry Kissinger's view of détente was not much different from Moscow's. It was also minimalist. His oft-quoted expostulation as to "What in God's name" was the meaning of "superiority" when neither side could escape from the mutual suicide pact implicit in MAD referred only to the apex of the strategic balance. He entertained no illusion that détente would in and of itself constrain Moscow's pursuit of Soviet state interests; nor did he intend to allow it to hamstring US policy objectives. He did see mutual benefit in the evolution of a web of interlocking interests, through the slow but steady proliferation of trade and other ties, and he clearly thought that the resultant state of interdependence might soothe rancor occasioned by differing national interests. Kissinger could scarcely be accused of presuming that conflicts would not yet occur; rather, he sought to establish a communications base and a pattern of accommodation which, though perhaps tentative and tenuous, might at least constrain the more dangerous oscillations of distrust and misperception, and hence condition greater receptiveness to possible negotiating formulas. Other American politicians did propagate more naively one-sided views of détente, and the acceptance accorded these views in turn fertilized the skepticism that was to be détente's undoing. The complexities of this process will be discussed shortly.

But first the topic of US military-political initiatives must be addressed. Barely a year after the signing of SALT I, Secretary of Defense James Schlesinger began developing what was to be called the Schlesinger Doctrine.[11] This "doctrine" was anchored in the accuracy claims made for America's new (MIRVed) Minuteman III missiles, and uncertainty as to whether the new volunteer forces would have a sufficient manpower pool to meet all contingencies. Schlesinger hypothesized that the United States might, for example, find itself unable to draw enough troops from Europe and other theaters to deter or counter the threat or actuality of a Soviet incursion into the Middle East. Inability to mount an effective conventional response would face Washington with the unpalatable choice of resigning and being defeated or resorting to strategic war. It was an unacceptable choice, especially since America's threat to bring about the Armageddon death-pact might well in the circumstances be seen as a bluff that could safely be called. There were, of course, alternate ways out of the dilemma. Manpower levels could be augmented through a reinstitution of the draft or through increased financial inducements to enlist. Responses in regions more favorable to American designs might alternatively prove effective. Schlesinger, however, chose to propound the idea that the way out of the impasse lay in the selective utilization of America's new missile prowess. American determination might be demonstrated through limited "surgical" strikes against remote Soviet installations (say, Siberian dams), targets carefully chosen to minimize fatalities and collateral damage.

Many commentators were perturbed at the very concept of using nuclear-armed strategic missiles for, in effect, tactical purposes. For one thing, it promised to lower and dilute the conventional-nuclear "threshold." And the associated fear that the genie, once let out of the bottle, would prove uncontrollable did indeed appear particularly apropos in this case.

There were two fatal flaws to Schlesinger's premise.[12] One derived from the fallacy of easy assumptions about first-strike eventualities. Quite apart from dubious suppositions about accuracy (extrapolated from testing-range environments that necessarily differed from those of real-life contingencies), the fact was that perfectly coordinated first-strike scenarios were possible only in theory. The practical problems of ensuring the simultaneous or near-simultaneous arrival on widely dispersed targets of such numbers of missiles from such widely dispersed launch locations as envisaged by these scenarios were nearly insurmountable, and even more so if one wished to allow also for the complex and up-to-the-last-minute checking procedures (and attendant delay possibilities) that the ideal called for, for each missile. The counterforce requirement of first-strike concepts would of course dictate last-minute target changes, whereas a retaliatory missile preprogrammed to strike cities or other fixed targets could dispense with time-consuming needs of final adjustments and calibration.

The point, however, is not just that a significant segment even of the opponent's land-based forces would survive long enough to perform a second-strike function. More importantly, the targeted nation's decision-makers could never feel confident that an initially limited number of incoming missiles were not but the forerunners of a larger strike force. A call from an American President might be less than comforting. In view of the perceived benefits of surprise, Moscow could scarcely expect Washington to have compunctions of honor, once a decision to attack had been reached, not with millions of American lives at stake. And if one were to suppress this injection of realism, one would still come face to face with a logic of equal horror. Rather than accept the diktat implied by an American strike at their own hinterland, however surgical and however limited, Soviet leaders would be compelled to show equal resolve; they could not afford the political penalty of not aiming a similar strike at the Tennessee Valley Authority (?) or . . . ? Ad infinitum. Neither side could afford at any time to show the self-constraint that the theory would require at each and every step. But even if one were to allow for the sake of argument that they could, neither one would be able to afford *not* to take the next step. The respective logics appear to lead inexorably to the same results. And one might note that there is not one sentence from the voluminous library of Soviet strategic writings to suggest that Moscow would even contemplate reticence or resignation in the face of a detonation on Soviet soil; very much to the contrary.

Soviet sources in fact dismissed selective targeting and surgical strike scenarios as they had once dismissed the concept of flexible response.[13] In each case Soviet commentators ridiculed the expectation that combat levels could be neatly categorized and controlled; the uncertainties of actual combat, especially nuclear, were seen to contain escalatory dynamics beyond the ken of theory. Moscow professed to see only one rationale for even talking in these terms. The rationale was said to lie in the fact that more cataclysmic images of nuclear war hampered the American military's obsessive search for ever-increased funding.

Disregarding the propaganda aspect (and possible relevance) of the final point, other elements of the undoubted continuance of postwar American strategic doctrine warrant comment. President Eisenhower's "massive retaliation" concept underscored American superiority during the 1950s, a time of manifest Soviet limitations. But the emergence of unquestioned Soviet ability to strike the US homeland occasioned European unease that Washington might no longer be willing to strike Soviet cities in retaliation for attacks on Paris, Bonn, or London, if such action meant the inevitable obliteration also of New York, Chicago, and other American centers. The Kennedy administration's doctrine of flexible response underscored the fact that the United States retained even greater superiority at the "theater-nuclear" level than it (still) did at the strategic level. The doctrine was supposed to demonstrate war-fighting capability at all conventional and battlefield or theater-nuclear levels; unquestioned ability to meet all kinds of escalation with superior force potential was supposed to deter escalation in the first place. The doctrine never quite fit the facts, since NATO never did deploy the additional conventional units that full coverage of conventional contingencies would have required. The inevitable consequence (and perhaps original intent) was that NATO focused on the upper rungs of the escalation ladder, where its superiority was unquestioned. NATO announced its readiness and determination to initiate the use of "tactical" nuclear weapons if required to halt enemy advances. There might still be doubt whether Moscow would be taken out in exchange for Paris. But at least now there was a more credible threat to Eastern European capitals, installations, and supply routes. As the 1970s wore on, however, the proliferation of Soviet theater-nuclear weapons eroded the Western edge in this area also. The Soviet ability to escalate (and match escalation) through the entire spectrum of "flexible response" began to look as potent as Washington's. Hence the Schlesinger Doctrine.

The Schlesinger Doctrine focused, as did its predecessors, on the area of greatest remaining US advantage. By drawing attention specifically to the most notable US edge of the 1970s—namely, the superior accuracy potential (and warhead numbers) of American strategic forces—it served to underscore a degree of continuing American preponderance. Psychologically it served to proclaim a degree of continuing American superiority. The outgoing Carter administration's embrace of Presidential Directive 59, or PD 59 (essentially a more elaborate version of the Schlesinger Doctrine), at the dawn of the 1980s, was presumably similarly motivated. As had been the case with its precursor, it was subject to attack, by Soviet and American critics alike, on the grounds that its very promulgation undermined the doomsday connotation of nuclear war and therefore inherently made the contemplation of such war more plausible, and also on the grounds that its implementation was fraught with such horrendous margins of uncertainty that no sane planner could feel real-life confidence in its primary control postulate.[14] But at a time of increased American and Allied public concern about the steadily accumulating strength and versatility of Soviet force elements, the publicity accorded PD 59 did serve to refocus attention on the fact that America's technological edge had not been negated.

The view that America's evolving strategic posture represented assertions of confidence and residual margins of superiority is suggestive. For one thing, it clarifies the perceptual parameters that allowed American defense planners to greet the limited drop in "real military expenditures" that occurred through the early and mid-1970s with such equanimity (although one must add, as another factor, that the pro forma decrease was in any case artificial, since it derived from a Vietnam-inflated base). The Reagan administration of the early 1980s was to look at the apparently coasting nature of military procurement patterns following the "bargaining-chip jump" of the SALT I negotiating era, and pronounce the judgment that planners of the day had been overly sanguine in the face of emerging Soviet prowess. But while using this argument to fuel a new dramatic spurt in military expenditures (indeed, it was to be the most precipitous funding increase in peacetime history), the administration appeared to see no contradiction in its accompanying and categorical assertion that American military procurement programs could, if necessary, be further intensified, to a point where Soviet emulation efforts would bankrupt the Soviet economy. The more eminent Western specialists on Soviet affairs greeted the assertion with extreme skepticism.[15] But the reality of the Reaganite extrapolation was perhaps less immediately relevant than its connotation of continuing and fundamental confidence.

If American confidence remained more assured throughout the decade of the 1970s than was sometimes presumed, then weight is also added to another argument of interest. A number of noted (anti-Soviet) Marxists saw détente and its economic corollary—Soviet willingness both to increase outside technology imports and to permit increased American financial participation in Soviet resource-extraction ventures—as certifying the failure of the Soviet economic model, a sign that Soviet leaders were resigning themselves to a degree of neocolonial dependence.[16] The interpretation echoed the obsessive fear that in another context drove the Maoists to oppose their "pragmatic" colleagues' espousal of increased Sino–American trade ties. The specter might well appear somewhat far-fetched in both cases. The fact is that Kissinger's quiet diplomacy pursuit of an expanding web of US–Soviet "linkages" did occasion Soviet concessions on a number of issues, most particularly emigration. The demise of the US–Soviet Trade Agreement in 1974, following Senator Henry Jackson's successful effort to make its congressional passage dependent on a public Soviet commitment to increase emigration levels further, was no surprise to Soviet specialists. The public nature of its connotation of blackmail could have been tolerated by no sovereign state. Predictably, emigration levels plummeted. But while the (perhaps calculated) failure of the Jackson maneuver testified to the limits of purported Soviet dependence, the success of Kissinger's behind-the-scenes tactics did nevertheless certify that Moscow was willing to (forced to?) pay at least a limited price for the benefits of détente.

Ironically, there is evidence that Soviet leaders themselves overestimated the economic benefits that could be derived from détente, and that they were consequently inclined to pay a higher political price than might have been warranted by more realistic expectations (although it was always clear that they could not afford to tolerate and concede to the extortionate demands of the Jacksonites). Moscow not only was willing to pay a price but may unwittingly have paid a considerably higher price than expected. The former point relates to the increasingly sluggish nature of the Soviet economy during the 1960s. There emerged a consensus that economic strategy stood at a new crossroads, the choice being to permit dramatic and fundamental reorganization in favor of greater emphasis on market mechanisms (the "Lieberman" alternative), or else to increase technological imports from the West in order to permit the centralized economic structure to meet the ever more complex challenges of an expanding and increasingly sophisticated economy. The latter course was deemed politically preferable. But while the Soviet economy continued to expand at a measured pace through the 1970s, the credit may have been due more to successful tinkering and improved indigenous means than to the expected panacea.

The level of Western imports never did exceed more than a fraction of a percent of Moscow's own investment efforts.[17] And although the high-technology content might have been presumed to give it disproportionate effect, this was largely offset by the evident fact that the imports proved more difficult to absorb than had been expected.[18] Rusting examples languishing on the docks of Odessa and elsewhere stood mute testimony to the problem. Another unforeseen consequence arising from officialdom's chosen policy lay in the fact that technology emulation was a lengthy process, so much so that the products that finally emerged tended to have been made obsolete by the intervening strides of Western technology. After all, imports embodied the results of older technologies, not ongoing innovation. It could be argued that mistaken Soviet faith in the benefice of détente was in fact the prime cause of continuing Soviet technological inadequacies, in that it appeared to lessen the imperative to pursue indigenous innovation. The failure was far from total. But Moscow had clearly not succeeded (at least not fully) in generating the type of developmental spurt that had accompanied Germany's and Japan's conscious switch from technology import and copy dependence.

A 1981 article in *The Economist* of London marveled at the success of a Soviet computer development program that had been sparked when President Carter vetoed a 1979 computer sale. The article noted that more severe Western trade restrictions might well spur rather than retard Soviet technological progress.[19] It was a point that ran totally counter to the arguments brought forth by those who wished to extend and tighten NATO's strategic goods embargo list. But it drew support from the stance of many of America's more prominent industrialists (symbolized by David Rockefeller and the Chase Manhattan Bank), who had been opposed to arguments and trends that restricted the flow of technological interchange: they had always appeared supremely confident; and *The Economist*'s example suggested they may have been prescient, not naive.

But by this time the last vestiges of détente appeared in any case to have been smothered by the arrival of Soviet troops in Kabul. The unraveling of détente did of course have deeper roots. 1974 saw the congressional spiking of the trade agreement negotiated two years before by Nixon and Brezhnev. The same year saw domestic political pressure force Nixon to defer the hoped-for signing of a follow-up SALT II agreement. A similar determination on the part of his successor, Gerald Ford, was to be similarly thwarted, in late 1974 (when a minimum agreement on US-Soviet "guidelines" for SALT II proved to be all that was politically possible) and finally in 1976 (when the pressures of a more conservative Reagan nominating challenge stayed his hand). The third consecutive American President to identify SALT II as essential to the furtherance of American interests, Jimmy Carter, finally did feel able to sign the document in 1979. But he could not secure congressional ratification.

Various sectors of the American polity had always opposed détente. Their reasons ranged from ideology to pecuniary and personal-professional bias (in the case of the military-industrial complex and its multitudinous dependents). But to understand the groundswell of urgency that finally allowed their arguments to prevail, one must perforce first review the facts and myths surrounding Moscow's three most controversial foreign policy initiatives of the 1970s.

Moscow's intervention in support of the MPLA during the Angola "civil war" of 1975 was to stand as the first successful example of Soviet military-political involvement in the affairs of an African nation.[20] It therefore acquired extraordinary political significance. The demonstration of will and newfound military muscle catapulted Moscow into the ranks of potentially dependable and able allies. The Soviets had never before dared to assert their interests so openly and so purposefully in a distant arena. They had never before possessed the physical wherewithal, in terms of long-range air and naval capabilities. Their Angolan venture testified to new potency and signaled the end of the era that had equated outside involvement in African affairs with Western financial, political, and military interests.

The time and location for this demonstration of resolve and steadfastness was of course judiciously chosen. The MPLA enjoyed extensive support among African governments. Its more narrowly tribal-based rivals were too obviously unrepresentative (they themselves acknowledged that they would need MPLA support if they were to govern effectively), and too tainted by their dependence on the Republic of South Africa for support, to secure the backing of a single black African regime. US Secretary of State Kissinger tried to secure authorization for US action against the MPLA and its allies. But as it became clear that even America's allies in black Africa felt forced to condone, if not support, a Soviet–Cuban intervention said to be motivated by the need to repulse South African aggression, Congress balked. When the head of the State Department's own Africa desk warned against the larger consequences of too intimate an association with Pretoria's apartheid government, and when associates of that government openly admitted that South African troops had indeed entered Angola before the arrival of the first Cuban contingents, the case was clinched. With Congress vetoing larger designs, the administration had only covert operations to fall back on. Predictably, these proved not to be enough. Two points are relevant: (1) Moscow could feel fairly confident that the political circumstances of the time would stymie Washington's ability to respond in kind; and (2) with non-native opposition whittled down to covert-size (mercenary) and South African units, the scale of operations required was well within the capabilities of Moscow's new force potential.

The Soviet Union's and Cuba's intervention on the side of Ethiopia in 1978, after Moscow's erstwhile ally Somalia chose to ignore Soviet prescriptions for the resolution of Somali–Ethiopian differences (through a larger Federation), was to prove equally fortuitous. Once again America's best friends in black Africa felt forced to condone the Soviet initiative. After all, Somalia was the undisputed aggressor. The US and other NATO countries had themselves repeatedly affirmed the view that the Ogaden region, the object of Somali action, was an integral part of Ethiopia. Angolan conditions had allowed Moscow to cloak its particular policy interest in the guise of anticolonialist, anti-imperialist and anti-apartheid moral imperatives. The Ogaden context similarly allowed Moscow to present its policy preference as a function of higher ideals. The injunction against forcible changes of established boundaries ranked high on the Organization of African Unity's list of moral absolutes. The ideals seen to have been involved in Angola may have been more emotionally charged. But the one called into play by Ethiopian defense efforts had practical urgency (since all African boundaries, legacies of colonial rule and considerations, dissected at least some tribal territories, each and every state felt potentially vulnerable to irredentist claims and aspirations).

Once again Kissinger attempted to orchestrate American aid to the side that *ipso facto* challenged Soviet purpose. As in the Angolan case, one presumes that he was aware that his awkward choice of allies sparked widespread alienation in Africa, alienation that provided grist for Moscow's propaganda mill. One must conclude that he was driven by unease lest the political-military impact of Soviet success prove even more detrimental to America's long-term interests. Once again, however, the US Congress proved more sensitive to politics of the moment and less apprehensive about future trends. Support for Mogadishu (Somalia) was restricted to that which could be supplied through covert means and operations, supplemented by inputs from the few states (Saudi Arabia and Egypt in particular) that professed to empathize with Kissinger's concern. The scale of opposition to Moscow's venture was still to be more substantial than it had been in Angola. Somalia's well-trained and equipped forces (the near-decade-long presence of Soviet advisors had left its mark!) were not as hamstrung by logistical and other considerations as Pretoria's army had been three years earlier. But Moscow's distant force potential had grown apace.

The Soviet Union's Angola operation had demonstrated an impressive ability to coordinate a multifaceted effort. Supplies had been brought in from a complex variety of embarkation points, ranging from Cuba, Iraq, and Algeria to Eastern Europe. But the restricted nature of the opposition had eased logistical air and sea requirements. And the weaponry involved had not needed to be of Soviet front-line quality. The Ethiopian contingency was considerably more demanding. The Soviets' ability to rise to the occasion testified to the rapidly improving nature of their distant force projection capabilities. It therefore imbedded and reinforced the Angolan lesson, that Moscow now could and would protect Third World allies.

A glance back through history may be apropos. The emergence of a new actor in the arena of world politics has always occasioned a period of acute strain. Established actors inevitably fear the extent, nature, and dynamism of the new aspirant's ambition. It necessarily challenges the old status quo, and hence also its implicit distribution of privilege and wealth. The new actor invariably espouses a form of manifest destiny (some version of which appears integral to all strains of ethnocentrism and chauvinism), as indignantly or assertively self-righteous as are those of the status quo power(s) of the moment. A tense time of testing, of the strength of residual power versus new license, ensues, sometimes resolved peacefully, sometimes not. The participants in these dramas have always felt compelled to portray the conflict as one of ideology, or morality challenged by immorality: Christianity (or Islam, or whatever religion was appropriate) battling heathen forces, "Popery" confronting "Proddies," Shiah muslims carrying the banner against Sunni deviants, ad infinitum. History would not have been surprised by the analogous inclinations of Soviet and American leaders.

The details of Soviet and Cuban action in Angola and Ethiopia were less relevant, and they were far from unique. The late 1970s witnessed numerous occasions of Western military-political assertion in Africa. French paratroopers, for example, played a decisive role in more capitals than did Cuba's expeditionary corps. And while France's role did not always match US interests, neither did Cuba's with those of Moscow. Cuba had its own rationale for action (African involvement echoed the peculiarities of Cuba's traditional self-image as a nation straddling both Latin and African cultures; it reflected the residual dynamism of Che's revolutionary-idealistic heritage; it gave purpose to that part of the educated elite that could not be absorbed by the socioeconomic infrastructure at home, and whose consequent disaffection might otherwise have been disruptive; and, to the degree that it served Soviet purposes also, it balanced a previously one-sided dependency relationship, securing for Castro a freedom to maneuver that he had not enjoyed since the early 1960s).[21] The Cuban engagements were more demanding of manpower, but then Western-installed regimes of the late 1970s did not face the type of outside pressure that South Africa continued to exert on Angola, and Somalia on Ethiopia. The point is not that the Soviet Union could call on more allies in more regions; it could not. It was the fact that the USSR could call on any allies, and sustain and support their involvement, that was novel, and arresting.

The Afghan imbroglio was in a different class.[22] More than any other event, it was to symbolize the end to the aspirations of the early 1970s. The decade had been ushered in on the crest of *ostpolitik,* détente, domestic prosperity, and relative international harmony. The 1980s opened with Soviet troops entrenched in Kabul, Sino–American support to anti-Soviet guerrillas, a world energy squeeze, recession, increased arms budgets, and international frigidity.

In a sense the Soviet resort to force in Afghanistan was but a reflection of the nasty but persistent tendency of great powers to ignore the niceties of international law whenever required by "national interests." 1979, the year that saw Soviet tanks rumble through Afghan towns, was replete with illustrations. China invaded Vietnam in February, with less success, but with a considerably larger force and perhaps greater brutality (if one accepts reports of the use of chemical and biological weapons). French and Moroccan troops toppled two African governments during that summer and fall. The United States could bask in the knowledge that it had desisted from the use of really large-scale military force since President Nixon's invasion of Cambodia. But it had, of course, built a rich tradition prior to that event. And it had supported subsequent allied operations.

The events in Afghanistan were also discouragingly typical, in that they mirrored an outside power's decision that further commitment was required to protect established interests from perceived jeopardy. Moscow's "presence" in northern Afghanistan had deep roots; after all, Britain's first invasion of the country in 1838 was designed precisely to check Russian influence.

Although proclaiming himself a Soviet ally, President Hafizullah Amin had ousted Soviet-supported Party members with ties to the clergy and other traditional sectors of Afghan society. He had also ousted Soviet advisors who preached caution, and he remained uncompromising in spite of alienation that generated rebel support in over 80 percent of the country. In view of evidence of limited Chinese and American support for rebel aspirations (and potentially for Amin), the brutal *realpolitik* ingrained in Russian perception left little leeway to opponents of action—especially once attempts at a more "discreet" coup had failed. To the Russian psyche, one China and one Ayatollah facing Moslem Central Asia were quite enough.

But as the Strategic Arms Limitation Treaty of 1972 had come to symbolize the gradual flowering of détente that had preceded it, so the Afghanistan crisis took on a meaning far beyond its immediate content: it capped and symbolized the unraveling of détente aspirations. The fact that the unraveling preceded the Soviet intervention ironically served to bolster advocates of intervention. Most of the possible disincentives to intervention had already fallen prey to the anti-Soviet mood of American domestic politics. NATO arms budgets had already been increased. Senate ratification of the SALT II treaty looked less and less likely. Hopes that Congress might yet sanction the trade agreement signed by Nixon in 1972 were dashed by the announcement that China's application for most-favored-nation status would be uncoupled from that of Moscow, and that China's would be granted (never mind that Beijing's human rights record appeared even more suspect than Moscow's).

American high-technology sales were already subject to increasing administration restriction and embargo. Soviet dissidents found disproportionate fame on American lecture podia and in the media. Andrew Young, the Carter administration's most prominent proponent of the thesis that all the world's ills might not be caused by Moscow but might owe something to local antagonisms, underdevelopment, or different forms of exploitation, resigned his UN ambassadorship. Early in 1979 Washington sent its first ambassador to Beijing—while invading Chinese troops remained on the soil of Moscow's Southeast Asian ally. As the date for Moscow's action drew near, the news media reported the imminence of the US Secretary of Defense's forthcoming visit to Beijing and American intentions to sponsor military-technological aid to, and defense policy coordination with, Beijing. The President seemed increasingly disinclined to accept the advice of the State Department and its Soviet specialists, choosing to rely more on his National Security Advisor, Zbigniew Brzezinski, a man whose views on occasion echoed the anti-Russian antipathies of his Polish ancestors.

Of course, Washington did not cause the Afghan crisis. The Soviets' "need" to invade Afghanistan was a function of their perversely paranoid concept of security, and of internal Afghan events. These factors would have "dictated" action even had the Shah remained on his neighboring Peacock throne. Similarly, it was not Western condemnation that deterred further advances. The last thing Moscow would have wanted was to send the Red Army into Pakistan. Its aim was much better served by American support for President Zia's unrepresentative dictatorship; it increased the likelihood that political and ethnic opposition to the Zia regime would acquire an anti-American hue. A future Pakistan, whether splintered or not, held definite promise from Moscow's vantage point. And even if such promise should be dashed, Moscow would be more likely to choose to encourage Indian intervention. Continued amity with India was far more vital in strategic terms than lingering aspirations for southern warm-water ports.

The most disturbing fact about "the new cold war" was that American anti-Sovietism had been built on a solid foundation of perceived Soviet perfidy, while Moscow's anti-American phobia rested on an equally convincing accumulation of perceived American villainy. Paranoia breeds paranoia. Distrust finds its own motivating rationale. It is worth remembering that as Zhdanov's demagoguery during the late 1940s had been seen to justify and perhaps necessitate a Joseph McCarthy, so McCarthy's vituperation had been seen to justify and necessitate Zhdanovite prejudices. As Soviet intervention in Afghanistan provided the clinching argument for American hardliners to rally majority support, so events preceding it had allowed Soviet hardliners to gain dominance in their policy councils.

The weapons procurement programs initiated by the Carter administration during its final years in office, before the Afghan eruption, provided graphic testimony to the nature and ramifications of emergent trends. One such program also illuminates the role played by misperception, real or feigned. It derived from the debate occasioned by Moscow's deployment of multiwarhead intermediate-range SS-20 missiles. To Moscow the SS-20 finally redressed the imbalance caused by the potency of America's forward-based systems (FBS) of carrier- and land-based aircraft with the range and sophistication to reach Soviet targets. Previous Soviet intermediate-range missiles had been both less accurate and more vulnerable (to hostile "take-out") than American systems. In an atmosphere of mutual suspicion, however, it was perhaps inevitable that NATO would ignore the fact that the SS-20 did not negate continued FBS efficacy and would focus instead on the indisputable point that SS-20 accuracy and survivability entailed a greater threat to NATO territory than had previously existed. Hence the phenomenon of NATO defense budget increases and the American decision to design and deploy a yet newer generation of theater-nuclear weapons in Europe, amid Moscow's protestations that these (and not the SS-20s) constituted the beginnings of a new arms race.

But President Carter's MX decision was perhaps even more illustrative. The program called for a gigantic complex of multiwarhead missiles darting among multiple shelters—the ultimate shell game. Its rationale rested on the calculation that the quantity and accuracy of Soviet missiles threatened the survivability of fixed-location missiles like the Minutemen.[23] To some, the logic of threat to silo-based missiles invited concentration rather on less vulnerable airborne and seaborne forces, forces that in and of themselves possessed awesome overkill (especially since the old objection, that their once inferior accuracy precluded certain targeting options, had been largely overtaken by technological advances), or on protected or mobile land alternatives—and these were later to come to the fore. But the possibility that the phasing out of older land-based force concepts might be ascribed to Soviet prowess still grated on too many sensibilities. Hence the MX system (the viability of which was predicated on the SALT II treaty and its restrictions on Soviet and American warhead numbers) was ironically pushed with all the more vigor once it became clear that SALT II would not receive congressional approval. The fact that a Moscow unrestrained by SALT obligations could deploy as many warheads as required to dispose of all the shells, filled or not, and the fact that further Soviet warhead increments might be procured more cost-effectively than any conceivable MX expansion scheme, appeared less relevant than the political imperative of facing down the Soviet challenge. Congressional willingness to fund the more than $60 billion MX system at a time of budgetary restraint and cutbacks in social services bespoke the depth and strength of this imperative.

The momentum of that same imperative was confirmed by the decision of the early Reagan administration to proceed with record increases in US defense outlays, in spite of calculations by some of America's most prominent economists that this would lead to deficits far greater than those generated by President Johnson's decision to simultaneously finance the Vietnam War effort and his domestic Great Society goals—deficits blamed for the inflationary spiral of the 1970s.[24] The administration obviously felt the need to be absolute. It was acutely discomfited by the political advantages that it thought had accrued to Moscow as a result of what was seen to have been successful Soviet efforts to pursue military-political objectives abroad. Washington bemoaned the fact that more substantive ripostes to Moscow's ventures in Angola and Ethiopia had not been possible. The accumulated impact of selective worst-case presentations of balance-of-power trends was also making itself felt.

The Soviets played their part in spurring the resurgence of American arms advocacies. They flaunted a silo reload capability, thus implying a further quantum leap in missile numbers.[25] And they gave prominent and emphatic media coverage to older, but only recently released American war-initiation contingency plans. The suggestive coverage appeared part of a wider effort to improve war preparedness. In Washington it led to renewed awareness of the Soviet doctrine's "war-fighting" predilection. There were those who empathized with the Soviet profession that the doctrinal stand reflected primary and necessary obsession with the ultimate fear, that it had no more conflict-initiating relevance than did America's chosen focus of "deterrence," and that it should in fact be seen as a more unequivocal substitute for "deterrence." One might add that the years of manifest Soviet strategic inferiority did make it more incumbent on Moscow to leave no doubt about its readiness actually to fight; assured of its preponderance, the United States could be more confident that it would not have to fight. The different conditioning process of Soviet history and memory should, of course, also be remembered. The point is simply that revived American attention to this nuance of Soviet doctrine was not conducive to the lessening of fear sparked by convictions of Soviet prowess.

There was one other factor that fueled Reagan's urgency, namely, the fear that a purportedly superior Moscow might realize the ephemeral nature of its "window of opportunity," and might therefore feel driven to take advantage of it. The CIA's upward revisions of the true costs of Soviet arms efforts suggested that Soviet investment in both research and procurement had significantly outstripped that of the United States. But CIA studies of the late 1970s also projected continuing economic problems for the Soviets, in particular a severe shortfall in their oil production in the late 1980s. The Soviet intervention in Afghanistan, though clearly motivated primarily by other considerations, served at the same time to renew appreciation of the proximity of Soviet borders to Middle Eastern oil fields. Washington became perturbed that the Soviet Union, at the apogee of its power, and cognizant of negative trends concerning both its domestic economic potential and the surrounding security environment (the firming up of a US–Chinese "alliance" and increased Japanese defense spending in particular), might feel tempted to preempt. At one extreme, the fear focused on the previously suggested hypothesis that Moscow might attempt to take out US land-based forces, according to the calculation that the consequently greater residual of Soviet force elements would suffice to deter US reprisal; Moscow might gamble that the inevitability of US city destruction in response to American retaliation might stay a President's hand and lead to resigned acceptance of Soviet preponderance. At the other extreme, there was fear that the US forces' acknowledged inability to match the Soviet force potential that could be brought to bear against Iranian oil fields (if only because of logistics and geographic dictates) might be interpreted as an irresistibly tempting vacuum by a Moscow confident of its larger force conglomerate's ability to deter escalation.

Yet all the pro forma premises, calculations, and fears of US policy appeared open to serious challenge. Although Moscow's overall force structure had seen tremendous improvement during the 1970s and had become numerically preponderant in a number of different categories (such as nuclear megatonnage and total men under arms), the United States retained a distinct advantage in perhaps the most important single quantitative indicator: deliverable warheads. And while the Soviet Union was in the process of developing accuracy technologies that could threaten American silos, the United States had been deploying similar technologies since the early 1970s, aboard the Minuteman III missiles. One might even argue that already existing US prowess in this regard dwarfed Soviet potentials, since a far larger proportion of Soviet strategic power continued to be deployed on land (79 percent of warheads, versus 21 percent in the United States). Notwithstanding unquestioned Soviet qualitative strides and the undeniable whittling away of America's technological lead, it is illustrative to note the congressional testimony of the Pentagon's chief scientist, W. J. Perry (discussing 1980 defense appropriations). Surveying the "Twenty Most Important Areas of Basic Technology," he asserted unchallenged US superiority in seven, continuing but diminishing US superiority in five, ascertainable US–Soviet parity in four, and presumed US–Soviet parity in the remaining four. Not in a single area did he concede Soviet advantage.[26] The summary obscured the dynamism of Moscow's improving prospects and the implications of Moscow's past success in innovatively optimizing dormant potentials of known technologies. Nevertheless, its significance remained real.

The specter of US land-based missile vulnerability was similarly met with shrugs by many observers, who considered land-based forces to be essentially redundant. Their argument noted the estimate by some of the world's most eminent nuclear scientists that existing arsenals already possessed perhaps 50 times the yield-effect potential needed to destroy the respective societies.[27] With 79 percent of US warheads (and 50 percent of US megatonnage) deployed on less vulnerable sea- and air-based delivery vehicles, the assuredly available redundancy of US retaliatory potential appeared of an order of magnitude that could not be dented by even the worst of worst-case assumptions. There was also the point that newer, more accurate technologies—especially aboard submarines, whose relative invulnerability allowed them the luxury of surfacing to doublecheck and fine-tune targeting coordinates—deprived land-based forces of their former monopoly of precision-targeting options.

Richard Garwin (scientific advisor to five Presidents, going back to Eisenhower) and others added that theoretical ballistic missile and warhead target accuracies were in any case deceiving.[28] Official data reflected the results obtained by calibrating gyros and accelerometers over lengthy periods of testing, over peacetime test ranges. Wartime trajectories would be quite different, involving gravitational and atmospheric phenomena that the instruments had not been programmed to counter. And potential satellite-provided data promised only partial recompense. The consequent discrepancy between theoretical and real accuracy might be minor, and would certainly not affect counter-city targeting, but it could well have a considerable effect on counter-silo designs. One might furthermore mention that analogous doubts attached to cruise missile potentials. Its terrain-hugging ("contour-matching") and terminal guidance appeared relatively effective in test locales in the southwestern United States, but liable to be thrown off track by forestation and snow—Russia's most prominent features. When put beside the acknowledged problem and effect of "fratricide," awareness of the imperfections and uncertainty of accuracy predictions sounded the logical death knell to first-strike fears—quite apart from the de facto impossibility of securing the exquisitely timed impact patterns that such specters assumed.

On the subject of redundancy, Lord Zuckerman (the former British government science advisor), Lord Mountbatten, and others made the point that just one of the now older Polaris-Poseidon–type submarines could deliver as much explosive power as was employed by all the combatants of the Second World War put together.[29] For insurance purposes, one might want two, five, or 10 . . . . Even that number was near-infinitesimal in the context of the arsenals of the early 1980s.

Then there were those whose skepticism was nurtured by earlier fears of Soviet advantage, by the image of mythical Soviet bomber squadrons of the 1950s and by the "Missile Gap Myth" of the early 1960s, and those who remembered President Eisenhower's somber warning about the dynamics and dangers that attend the ensconcement of a military-industrial complex. In the words of perhaps the most prominent of American Sovietologists, Ambassador George Kennan:

> The present Soviet and American arsenals . . . are simply fantastically redundant to the purpose in question . . . something well less than twenty percent of these stocks would surely suffice for the most sanguine concepts of deterrence . . .
>
> . . . They have, of course, their share of the blame. . . . But we must remember that it has been we Americans who, at almost every step of the road, have taken the lead in the development of this sort of weaponry. It was we who first produced and tested such a device; we who were the first to raise its destructiveness to a new level with the hydrogen bomb; we who introduced the multiple warhead; we who have declined every proposal for the renunciation of the principle of "first use"; and we alone, so help us God, who have used the weapon in anger against others, and against tens of thousands of helpless non-combatants at that.[30]

CIA data of the late 1970s were subject to much criticism. There was no question that Soviet defense budgets continued to grow throughout the decade of the 1970s, although the growth of the Soviet GNP may have been commensurate; in other words, the relative burden on the economy may have remained fairly steady. The dispute revolved rather around the politically appointed B-team's report, which doubled the CIA analysts' estimate of just what that real burden was.[31] The doubling became the base for subsequent and current calculations. It was responsible for the figures that showed absolute Soviet outlays to be greater than those of the United States, and the relative burden on the Soviet economy to be twice as heavy as it was in the United States.[32] The image was obviously a crucial propellant to American fears.

What is less appreciated is the fact that the B-team did not double American estimates of the size and potency of the Soviet armed forces; they did not find new and previously overlooked Soviet armies, navies, strategic rocket troops, and the like. They only doubled estimates of the relative cost of existing forces. The relative cost to the Soviet economy remained unchanged. It was only the American appreciation of what this cost represented that was revised so sharply upward. In effect, the new estimates declared the Soviet military economy to be only half as efficient as previously thought, scarcely cause for sudden alarm.

All the CIA estimates, including the earlier ones that presumed greater Soviet military economic efficiency, have furthermore been attacked on methodological grounds. The CIA's methodology, especially that for aggregating economic indices and categories, is said to contain an inherent and inevitable upward bias.[33] If the powerful arguments to this effect are indeed correct, then a certain downward revision of one's estimate of the Soviet defense burden becomes mandatory—whether one concurs with the B-team's view of Soviet inefficiencies or not.

Finally, as concerns the CIA's projection of significant Soviet energy shortfalls during the 1980s, one must note that subsequent agency reports were far less alarmist. (There are those who feel that while Soviet leaders might ignore their own economists' pessimistic prognoses, they do not belittle CIA expertise; it is suggested that the original CIA report may in fact have been a decisive spur to the program of energy mix shift, extraction improvements, invigorated exploration, and conservation that now appears to have eased the situation.)[34] The point of the matter relates to the contention that the Soviet economy was so severely strained by the existing defense burden that it would not be able to match a new surge in American defense spending. The suggestion that a great increase in US outlays might literally break Moscow's ability to compete found frequent expression among the Republican election entourage of 1980. Skeptics pointed out that Moscow had sustained far greater efforts on far shakier economic grounds in decades past. Yet there was the riposte that Soviet economic growth was stagnating by the late 1970s even with still buoyant energy production; the presumed imminent loss of this latter cushion conjured up the possibility of prolonged economic crisis. But the picture looked very different if one proceeded from less extreme assumptions about the relative Soviet defense burden, if one acknowledged that the 1979–1982 nadir of Soviet economic growth was affected at least somewhat by the aberrational shock of three successive abominable harvests, and if one conceded a less troublesome Soviet energy outlook. Economic problems and perhaps sometime suffering, yes, but neither Soviet nor Russian history provided much support for the expectation or hope that Moscow's defense commitment would abate.

Certain administration supporters insisted on exaggerating their concerns to such an extent that they repelled potential sympathizers. Statements that the Soviet Navy was now twice the size of the US Navy was a case in point. One does not lump in-shore, coastal vessels with battleships, cruisers, and aircraft carriers.[35] The United States retained nearly a two-to-one advantage in tonnage capable of engaging in distant operations. Moscow had acquired significant long-range naval interventionary potential, which it had not possessed 10 years previously. But US aircraft carriers alone still contained more interventionary potency than the Soviet Navy could muster. Unfortunately, such overstatements and the responses they engendered merely clouded the issue. In most cases (short of Armageddon), the Soviet Navy did not need equality. The Soviet interventions in Angola and Ethiopia are illustrative. In both cases Washington could have brought greater power to bear. The United States was deterred not by Soviet military prowess but by political constraints. Future situations are likely to be similar. Ability to use and manipulate political currents and the correlation of local (native) force elements are likely to prove weightier on the scales than any crude balance of naval force alone.[36]

The same is true with statements blaming Moscow for all the world's terrorism. At a time when the pages of the Miami *Herald* were full of reports of anti-Castro and other self-styled guerrilla (terrorist?) groups acquiring combat training in the Everglades, at a time when the Catholic Church, Amnesty International, and other organizations documented indigenous state and other terrrorism in Latin America, the Middle East, and elsewhere—at a time, indeed, when the CIA itself was in the process of finalizing a study tracing the variety and complexity of terrorist support and inspiration[37]—the discrepancy between assertion and reality was too blatant. There was no question that Moscow had in the past embraced and pursued nonconventional strategies (ranging from ideological subversion, through industrial espionage, to instances of sabotage and, *in extremis*, terrorism), as part and parcel of the deliberate manufacture of an image of deterrence, and hence a freedom to maneuver, that was as yet far out of proportion to its real strength. The current status of this unorthodox component to Moscow's military-political stance clearly warranted attention. But ludicrous exaggerations merely invited mockery, diverting attention from the real problem(s).

The image of unremitting Soviet success in expanding influence and empire was equally unfortunate. Yes, apparent success in Angola, Ethiopia, and South Yemen did give Soviet power a global credibility that it had previously lacked, and the supporting role played by Cuba and, to a lesser extent, other allies again pointed to the fact that Soviet power projection was not one-dimensional. Yet Moscow would surely have been happy to give up all the gains of the 1970s if it could undo the split with China. . . . And then there is the long list of others who "got away": Egypt, the Sudan, Somalia, Guinea, Iraq, Chile(?), North Korea, Rumania(?), Poland(?), and so on. Many of the supposed victories were brittle. Angola, for example, had jailed or expelled the most pro-Soviet elements within the ruling MPLA, perpetuated Gulf Oil's monopoly over oil production, and extended its hospitality to Western multinationals also to other areas of industrial development; in

Ethiopia Moscow's oft-expressed preference for a ruling party organized along traditional lines remained unsatisfied.

By 1980 there could be no doubt that US military power suffered from a number of flaws.[38] Purported deficiencies in the strategic arena were less credible. The fact that 90 percent of the Reagan administration's dramatic defense budget boost was allocated to the conventional sphere indicated that even the new team was inclined to accept this point, in private if not in public (this inference was supported by evidence of renewed skepticism toward the gigantic MX program, skepticism that culminated in the October 1981 decision to restrict initial deployment to existing silos—which were to be "superhardened"—while devoting further research into land-mobile, air-based, and other deployment modes). The real problems revolved around the volunteer forces' inability to attract sufficient numbers and quality. The shortage of mid-level specialists, midshipmen, mechanics, engineers, and the like led to ships not being able to put to sea, restrictions on flying hours, and other encumbrances. Little thought had been given to the civilian economy's ability to retool for military emergencies, and its ability to do so at short notice had apparently been impaired. As concerned interventionary prospects, there was the fact that the US Army did not have a canteen large enough to hold the minimum daily water ration required for areas like the Middle East, and the fact that new tanks were "so big and heavy that only one tank at a time can be transported by the Air Force's biggest plane, the C-5A; at the moment the Pentagon wants 7000 M-1 tanks, but the United States has only 77 C-5A planes."[39]

Yet such problems and anomalies were self-made, and not the result of Soviet scheming. The Soviet Armed Forces' organization and practice suffer analogous and equally glaring contrasts between theory and reality.[40] The Soviet bureaucratic process throws up its own gremlins of mismanagement, bottlenecks, and short circuits. Other US problems find their counterpart in the combination of Siberian isolation, boredom, and Vodka. And the trends toward disproportionate minority representation among US ranks are mirrored by the demographically ordained "Asianization" of Soviet forces.

The only real uncertainty in the strategic arena lay in questionable prospects for decisive breakthroughs in laser and particle-beam technologies, breakthroughs that might allow for truly effective ballistic missile defense systems (BMDs).[41] Either side could deploy a million additional cruise missiles or MXs, yet this would not negate the other's ability to effectively obliterate the aggressor's society. Revolutionary BMD concepts were in a different league, since they struck at the heart of the mutual assured destruction (MAD) core of the present balance.

Most scientists remained highly dubious. Nevertheless, this was the one strategic domain where advocates of increased contingency research spending could bring compelling logic to their cause. Conversely, of course, the extreme danger to each of the possibility that the other might succeed or appear to succeed in effecting unilateral advantage could be seen as a powerful argument for arms control negotiations of substance.

So why the extremism of the Reagan administration's posture, if so many of the military and military-political arguments upon which it rested are so open to challenge? The answer appeared to be psychological, the perceived need to counter the image of rampant and unchecked Soviet imperialism. For though the real story of Soviet power projection was far more checkered, and though the real potency of US power remained awesome, the illusion of Soviet invincibility and US emasculation did appear to be widely accepted. But again the point must be made that the wound was largely self-inflicted. The credit for the widespread currency enjoyed by these flawed myths was one that the Soviet propaganda apparatus shared with Senator Jackson and other American opinion-makers, who for their own domestic political purposes had chosen to insist, ever since 1972, that Soviet power was overtaking US capabilities. It is doubtful whether so many Third World audiences, audiences socialized to automatic assumptions of British, French, and American preeminence, would have conceived of Soviet superiority had not a bevy of American senators and politicians tutored them to that effect. Nevertheless, if one accepted that the damage had been done, and if one believed that it really mattered, then one was also compelled to sympathize with the administration's belief that a major show of machismo and determination was necessary, not on military grounds but for psychological reasons, to counter false assumptions about US strength and American will. The fact that Reagan's budgetary bonanza was undiscriminatory, spreading equal largesse to all three branches of the military and their components, and not focused on specific requirements tailored to meet particular "threats" (these were subjects for later deliberations), provided further testimony to the thesis that the impetus was psychological.

But the truly staggering nature of American defense budget increases appeared outrageously out of proportion to the purely psychological requirement, especially if one considers the incalculably cheaper alternative of greater publicity for arguments such as Lord Zuckerman's; the type of public relations effort that could portray the opponent's capabilities as a mortal threat could just as easily portray it as ridiculous waste. But such thoughts were stillborn in the psychosis of the early 1980s. Washington pushed ahead with a version of the MX, two new strategic bombers, new Trident SLBMs and aircraft carriers for the navy, new "theater-nuclear" systems, and surging funding of a motley variety of BMD and antisatellite concepts—the undiscriminating largesse that attended the yanking of defense outlays from $144 *billion* in 1981 to a proposed $343 *billion* in 1986. (The MX scheme alone brought contract work, profits, and voting impact to an astounding number of congressional districts.)

And yet, not a single one of these programs could realistically be seen to hold out hope for the attainment of decisive American superiority. Even in the one sphere of uncertainty, that of ballistic missile defense, the odds against imminent success (of the kind that might really stymie the main opponent's penetration prospects) remained nearly insuperable. To be effective, space-based laser BMD would require an increase in laser brightness of at *least* a factor 10^6(!)over that which ground-based experiments had demonstrated by 1981.[42] And even if this quantum leap in technological capability could be effected, the system might be defeated by relatively simple maneuvers (for example, if the opponent attached reflective aluminum foil screens to its booster rockets) and would itself be vulnerable to nuclear attack. Anti-satellite schemes appeared much more feasible (though the highly touted space shuttle would be far too expensive to risk against a possibly booby-trapped satellite). But the range of defense mechanisms available, from static metallic foil deflectors to a series of active counter-anti-satellite possibilities, meant that pursuit of progress in this arena held scant prospect of superiority—only an overwhelming likelihood of accelerating effort and ever more dangerous instability.

Still, Moscow appeared impaled on the horns of the same precipitous logic. There was no letup in Soviet modernization programs. The even larger Typhoon submarines began to supplement already deployed Deltas. The upgrading of Soviet "theater-nuclear" and force projection means continued apace. They even began to build on their earlier experiments and experiences in the field of mobile missile systems. For a decade, throughout the 1970s, Moscow had exhibited extraordinary equanimity in the face of pro forma US accuracy potentials. The Soviets had not been panicked into abandoning their basic reliance on fixed land-based systems. In fact, their espousal of and preparation for naval "withholding" had confirmed that they remained confident that the larger portion of these land forces would survive and that naval means would therefore not be required for immediate retaliatory purposes. As indicated before, all dispassionate analyses supported the Soviets' confidence on this score. If they now really were in the process of developing their own MX, it was less a vindication of the concept than it was evidence of Soviet susceptibility to the powerful psychology of accumulating American argument. If Washington proclaimed newer strategic systems to be decisive, then political imperatives dictated that Moscow adopt analogous capabilities.

Three concluding thoughts suggest themselves. One relates to the obvious point that serious arms control was off the political agenda—the Reagan administration was quite explicit in its contempt for past efforts in the field. Its contempt was summed up in 1982, when President Reagan resurrected the canard that Moscow's yield advantage translated into real superiority—disregarding the fact that the United States had chosen not to go to the heavy missile route because its superior accuracy technologies allowed it to pursue a different option (an option that was more cost-effective, and that had a far greater impact on yield effect), and disregarding also the fact that US accuracy technologies continued to outpace those of the USSR (Soviet achievements of the early 1980s only equaled American levels of the mid-1970s), hence putting additional premium on America's remaining and significant lead in warhead numbers. What was even more perturbing was the fact that the systems now being developed promised to defy future arms control endeavors also. The new systems, with their emphasis on possible mobility (MX), concealment and dispersal (smaller cruise missiles), and defense concepts of uncertain efficacy (systems that by definition gave license once more to grotesquely exaggerated "worst-case" dynamics), were far less susceptible to monitoring than the ones they were replacing.[43] The systems were designed, *ipso facto*, to maximize the practical problems of realistic arms control.

The corollary prospect of impaired communication was itself unsettling. America's missile warning detection systems had just experienced 147 malfunctions over an 18-month period.[44] Each consequent alert was aborted at an early stage, and the evident quality of backup facilities soothed most observers. But, in fact, Soviet authors wrote of the need for "launch on warning" capability and suggested that Moscow might (if need be) adopt "launch on warning" as a strategy, to secure against future surprise decimation of the Soviet land potential. Something approaching "launch on warning" was also the *sine qua non* of continued American investment in improved strategic land forces. The minimum flight time of hostile missiles does not allow for the luxury of much second thought if it is believed that retaliatory force elements may be destroyed. The danger of accidental war appeared minimal. Yet to those skeptical of the infallibility of computers (and personnel), any unnecessary danger loomed as an unconscionable price to pay for deployment modes whose military rationale had become suspect, and whose continued survival seemed increasingly a function of political prejudice. Without substantive East–West dialogue, the problem could not be properly addressed.

The second point to be made refers back to the earlier evidence of Reagan confidence—notwithstanding the assumed rhetoric of vulnerability. The confidence found many expressions. The Reagan administration discontinued the limited grain and technology embargoes instituted by President Carter in the aftermath of Moscow's Afghan incursion, once it became clear not only that Moscow was securing offsetting supplies from elsewhere but also that the alternate suppliers might permanently enhance their Soviet market access at the expense of American business interests. But Reagan perpetuated Carter's other initiative, the extension of an American security blanket to Middle Eastern nations. This particular doctrine had been a unilateral promulgation assuming a right. It was not, like NATO, a response to friendly governments who perceived a threat, offered bases, and asked for alliance. No country in the region had extended base rights, and no country had requested formal alliance (although at least in some cases this reflected the corrosive impact of continued US support for Israel's occupation of conquered Arab lands rather than empathy for Moscow).

European commentators focused their fear on the possible practical consequence of the doctrine, especially in the context of Presidential Directive 59; the relative inferiority of the conventional forces that Washington could bring to bear in this particular area implied that American action against Soviet forces would be nuclear. But the nature of the promulgation was equally illuminating. In the unilateralism of its assertion it echoed the Brezhnev doctrine vis-à-vis Eastern Europe. Yet the Middle East was of course geographically close to being in Moscow's back yard. The point is that similar assertiveness by Moscow in an area likewise favorable to American power projection—say, the Caribbean as a whole—would have been inconceivable. A Moscow commitment to action against American interventionary designs in Central America would just not be credible, at least not outside the isolated context of Cuba.

The implied disequilibrium is confirmed by a consideration of the debate about theater-nuclear modernization programs in Europe. New Soviet systems served to solidify a long-established potential threat against America's allies; Washington's newer systems could strike at the Soviet heartland, not merely at Moscow's allies. To the extent that emerging American theater-nuclear capabilities were designed and deployed to constitute a de facto addition to the threat posed by strategic forces, they echoed the core issue of the 1962 Cuban crisis. Moscow had then been forced to desist from similar deployment; the American initiative two decades later challenged the reciprocity implied by President Kennedy's withdrawal of missiles from Turkey. The Pershing II and cruise missiles scheduled by Washington for European deployment during the early 1980s were similar to SS-20s on Cuban soil—yet SS-20 deployment remained restricted to Moscow's Eurasian rimlands. President Reagan's offer, on 18 November 1981, to defer Pershing II and cruise deployment if Moscow dismantled all its Europe-targeted missiles, old and new, testified to the noncomparability of the systems. The deal would leave some 7000 American nuclear warheads deployed in Europe intact, faced by a now emasculated Soviet counter; Moscow was apparently presumed to be willing to pay a disproportionate price for NATO "restraint." It might also be relevant to note that the vaunted accuracy of the SS-20, calculated by the US to have a circular error probable of 1500 feet, contrasted with a Pershing II CEP of 120 feet. (Translated, this means that the Pershing II's kill probability against hardened targets was nearly six times greater than that of the SS-20).[45]

The final observation relates to the relevance of superpower status in the 1980s. The two superpowers were far more uniquely "super" in their military capabilities than ever before. No other power, not even China, could remotely hope to present a real military challenge to either of them before the end of the century. But their massive Armageddon potentials, and their redundant abilities to "stir the rubble" to their hearts' content, appeared inversely proportional to their ability to dictate events. The ideological pretensions of both superpowers, once so potent, now lay tattered.[46] And, as noted elsewhere, the worldwide proliferation of armaments, and the mushrooming of Third World arsenals in particular, promised to make future interventions much more problematic. The superpowers retained obliteration as an option. But low-cost options looked fewer and further between.

The superpowers were not only finding it increasingly difficult to translate their elevated power potential into real influence or control in distant areas. At the apparent height of their military supremacy vis-à-vis the rest of the world, they were finding it ever more problematic to retain influence even over traditional allies.[47] France and other NATO allies joined Mexico in opposing America's military intervention to prop up the Salvadorian junta. Warsaw Pact members were less than enthusiastic about Moscow's Afghan policies. And Poland the ally had clearly become Poland the albatross.

The first cold war had seen the power blocs become rigid. As Reagan led the way to the second, and Moscow rose to the bait, the foundations of both blocs began to shake. Washington's and Moscow's ability to inflict the ultimate penalty was greater than ever, but the obvious

	First in service	Range (nm)	Payload	1972	1973	1974	1975	1976	1977	1978	1979	1980	1981
Delivery vehicles													
Strategic bombers													
USA B-52 C/D/E/F	1956	10 000	27 000 kg	149	149	116	99	83	83	83	83	83	83
B-52 G/H	1959	10 900	34 000 kg	281	281	274	270	265	265	265	265	265	265
(FB-111	1970	3 300	17 000 kg	66	66	66	66	66	66	66	66	65	64)
USSR Mya-4 'Bison'	1955	5 300	9 000 kg	56	56	56	56	56	56	56	56	56	56
Tu-95 'Bear'	1956	6 800	18 000 kg	100	100	100	100	100	100	100	100	100	100
(Tu-22M 'Backfire'	1975	4 000	9 000 kg	–	–	–	12	24	36	48	60	72	84)
			Long-range bomber total: USA	**430**	**430**	**390**	**369**	**348**	**348**	**348**	**348**	**348**	**348**
			USSR	**156**	**156**	**156**	**156**	**156**	**156**	**156**	**156**	**156**	**156**
Submarines, ballistic missile-equipped, nuclear-powered (SSBNs)													
USA With Polaris A-2	1962	n.a.	16 × A-2	8	8	6	3	–	–	–	–	–	–
With Polaris A-3	1964	n.a.	16 × A-3	21	13	13	13	13	11	10	10	5	5
With Poseidon C-3 conv.	1970	n.a.	16 × C-3	12	20	22	25	28	30	31	31	25	20
With Trident C-4 conv.	1979	n.a.	16 × C-4	–	–	–	–	–	–	–	–	6	11
With Trident C-4	1980	n.a.	24 × C-4	–	–	–	–	–	–	–	–	–	1
USSR 'Hotel II' conv.	1963	n.a.	3 × 'SS-N-5'	7	7	7	7	7	7	7	7	6	6
'Hotel III' conv.	1967	n.a.	6 × 'SS-N-6'	1	1	1	1	1	1	1	1	1	1
'Yankee'	1968	n.a.	16 × 'SS-N-6'	27	33	33	33	33	33	33	33	29	27
'Yankee II'	1974	n.a.	12 × 'SS-NX-17'	–	–	1	1	1	1	1	1	1	1
'Golf IV' conv.	1972	n.a.	4 × 'SS-N-8'	1	1	1	1	1	1	1	1	1	1
'Hotel IV' conv.	1972	n.a.	6 × 'SS-N-8'	1	1	1	1	1	1	1	1	1	1
'Delta I'	1973	n.a.	12 × 'SS-N-8'	–	1	7	12	18	18	18	18	18	18
'Delta II'	1977	n.a.	16 × 'SS-N-8'	–	–	–	–	–	4	4	4	4	4
'Delta III'	1978	n.a.	16 × 'SS-N-18'	–	–	–	–	–	–	2	4	10	12
			Submarine total: USA	**41**	**41**	**41**	**41**	**41**	**41**	**41**	**41**	**36**	**37**
			USSR	**37**	**44**	**51**	**56**	**62**	**66**	**68**	**70**	**71**	**71**
			Modern subs: USSR	**27**	**34**	**41**	**46**	**52**	**56**	**58**	**60**	**62**	**62**
SLBM (Submarine-launched ballistic missile) launchers on SSBNs													
USA Polaris A-2	1962	1 500	1 × 1 Mt	128	128	96	48	–	–	–	–	–	–
Polaris A-3	1964	2 500	3 × 200 kt (MRV)	336	208	208	208	208	176	160	160	80	80
Poseidon C-3	1970	2 500	10 × 40 kt (MIRV)	192	320	352	400	448	480	496	496	400	320
Trident C-4	1979	4 000	8 × 100 kt (MIRV)	–	–	–	–	–	–	–	–	96	200

USSR	SS-N-5	1963	700	1 × 1 Mt	21	21	21	21	21	21	21	21	18	18
	'SS-N-6 mod. 1'	1968	1 300	1 × 1 Mt	438	534								
	'SS-N-6 mod. 2' conv.	1973	1 600	1 × 1 Mt	–	–	534	534	534	534	534	534	470	438
	'SS-N-6 mod. 3' conv.	1973	1 600	2 × 200 kt (MRV)	–	–								
	'SS-N-8'	1973	4 300	1 × 1 Mt	10	22	94	154	226	290	290	290	290	290
	'SS-NX-17'	n.a.	. .	1 × 1 Mt (MIRV-cap.)	–	–	12	12	12	12	12	12	12	12
	'SS-N-18'	n.a.	4 050	3 × 200 kt (MIRV)	–	–	–	–	–	–	32	64	160	192
				SLBM launcher total: USA	**656**	**656**	**656**	**656**	**656**	**656**	**656**	**656**	**576**	**600**
				USSR	**469**	**577**	**661**	**721**	**793**	**857**	**889**	**921**	**950**	**950**
ICBMs (*Intercontinental ballistic missiles*)														
USA	Titan II	1963	6 300	1 × 10 Mt	54	54	54	54	54	54	54	53	52	52
	Minuteman I	1963	6 500	1 × 1 Mt	290	190	100	–	–	–	–	–	–	–
	Minuteman II	1966	7 000	1 × 1.5 Mt	500	500	500	450	450	450	450	450	450	450
	Minuteman III conv.	1970	7 000	3 × 170 kt (MIRV)	210	310	400	550	550	550	550	550	550	450
	Minuteman III impr.	1979	7 000	3 × 350 kt (MIRV)	–	–	–	–	–	–	–	–	–	100
USSR	'SS-7 Saddler'	1962	6 000	1 × 5 Mt	190	190	190	190	130	30	2	–	–	–
	'SS-8 Sasin'	1963	6 000	1 × 5 Mt	19	19	19	19	19	19	–	–	–	–
	'SS-9 Scarp'	1966	6 500	1 × 10–20 Mt	288	288	288	288	248	188	128	68	–	–
	'SS-11 mod. 1'	1966	5 700	1 × 1 Mt	970									
	'SS-11 mod. 2' conv.	1973	. .	1 × 1 Mt	–	970	970	970	890	800	690	580	520	460
	'SS-11 mod. 3' conv.	1973	. .	3 × 200 kt (MRV)	–									
	'SS-13 Savage'	1969	4 400	1 × 1 Mt	60	60	60	60	60	60	60	60	60	60
	'SS-11 mod. 3'	1973	. .	3 × 200 kt (MRV)	–	20	40	60	60	60	60	60	60	60
	'SS-18 mod. 1/mod. 3'	1976	5 500	1 × 10–20 Mt	–	–	–	–	60	120	180	240	308	308
	'SS-18 mod. 2' conv.	1977	. .	8 × 500 kt (MIRV)	–	–	–	–						
	'SS-19' conv.	1976	5 000	6 × 500 kt (MIRV)	–	–	–	–	80	120	180	240	300	360
	'SS-17' conv.	1977	. .	4 × 500 kt (MIRV)	–	–	–	–	–	50	100	150	150	150
				ICBM total: USA	**1 054**	**1 054**	**1 054**	**1 054**	**1 054**	**1 054**	**1 054**	**1 053**	**1 052**	**1 052**
				USSR	**1 527**	**1 547**	**1 567**	**1 587**	**1 547**	**1 447**	**1 400**	**1 398**	**1 398**	**1 398**
				Total, long-range bombers and missiles: USA	**2 140**	**2 140**	**2 100**	**2 079**	**2 058**	**2 058**	**2 058**	**2 057**	**1 976**	**2 000**
				USSR	**2 152**	**2 280**	**2 384**	**2 464**	**2 496**	**2 460**	**2 445**	**2 475**	**2 504**	**2 504**
Nuclear warheads														
Independently targetable warheads on missiles: USA					3 858	5 210	5 678	6 410	6 842	7 130	7 274	7 273	7 000	7 032
USSR					1 996	2 124	2 228	2 308	3 160	3 894	4 393	4 937	5 920	6 848
Total warheads on bombers and missiles, official US estimates: USA					5 700	6 784	7 650	8 500	8 400	8 500	9 000	9 200*	9 200*	9 000*
USSR					2 500	2 200	2 500	2 500	3 300	4 000	4 500	5 000*	6 000*	7 000*

* 1 January.

Source: Stockholm International Peace Research Institute, *World Armaments and Disarmament: SIPRI Yearbook 1981*, Taylor and Francis, London, and Oelgeschlager, Gunn & Hain, Cambridge, Mass., 1981, pp. 274–275. Copyright © 1981 by SIPRI.

political cost and the cost in terms of military-escalatory dangers had begun to look exorbitant. In such a context, the drastically diminished nature of their economic control options acquired special importance. In 1981 American economic blackmail against Nicaragua failed because Managua was able to turn to Libya, Mexico, and Western Europe. By 1981 Moscow's financial sway over Poland was diluted by Warsaw's ability to turn to Berlin and the International Monetary Fund. And even where Washington and Moscow might still choose to impose their policy preferences, even in those areas where they might confidently defy the political and military-escalatory costs, there remained the fact of ever starker economic costs. Their economies were so strained that the thought of adding more economically dependent and draining vassals was itself a deterrent that could not be lightly dismissed. By 1981 there were many who thought that the immediate economic penalties of reestablishing total control in, for example, Poland and El Salvador constituted the single strongest deterrent to all-out Soviet and American action. Soviet consumers might willingly tighten belts if necessary to confront American "war-mongering," but they were not keen to subsidize "ingrate Poles"; US citizens might accept social service cutbacks said to be necessary to fund the additional defense increments supposedly required by the new specter of Soviet threat, but the middle class was less inclined to accept sacrifices for the sake of a Latin junta.

Epilogue. The spring of 1982 brought announced Soviet willingness to halt SS-20 deployment, a much publicized strategic arms freeze proposal sponsored by Senators Kennedy and Hatfield, a renewed upsurge of the "European Nuclear Disarmament" movement, and the sudden emergence of antinuclear sentiments as a potent political force in the US also. By May the Reagan administration felt compelled to call for the start of "strategic arms reduction talks" (START). But though it appeared that talks might indeed begin, the nature of the proposal allowed little ground for optimism. Reagan proposed ignoring bombers and cruise missiles, areas of marked US advantage; reductions were to focus on missiles only, and in particular on land-based rockets (on which Moscow was disproportionately dependent). Democratic party spokesman Edmund Muskie commented that the proposal appeared designed to suggest willingness to negotiate, yet ensure negotiation failure; two days later Secretary of State Haig confirmed that the US arms buildup would not slacken.[48]

NOTES

1. The evolution in Soviet strategy and deployment during the 1970s is analyzed in C. G. Jacobsen, *Soviet Strategic Initiatives: Challenge and Response*, Praeger Publishers, New York, 1979; 2nd ed., 1981.
2. The emergence of a new branch of Soviet strategic theory, dealing with interventionary-type prospects and requirements, was first traced in detail (by this author) in a U.S. government-commissioned study; most of the data therein presented, plus later Soviet pronouncements of relevance, may be found in Jacobsen, *Soviet Strategic Initiatives*, Chapters 2–4.
3. V. M. Kulish, *Military Force in International Relations*, JPRS translation 58947, May 1973, p. 103.
4. J. McConnell, "Soviet Coercive Naval Diplomacy," quoted in Jacobsen, *Soviet Strategic Initiatives*, p. 16.
5. *Voprosi Istorii KPSS*, May 1974; FBIS translation, 30 May 1974.
6. L. Brezhnev, *Report to the Congress*, Novosti, 1976, p. 12.
7. B. Pyadyshev, *USSR–USA: Confrontation or Normalization of Relations?* Novosti, Moscow, 1977, p. 99.
8. V. Matveev, in *Izvestia*, as quoted by W. Inglee, *Soviet Policy Towards Angola*, Library of Congress Research Service, Washington, D.C., 1977.
9. M. Suslov, *Kommunist*, 21 July 1975.
10. Editorial, *New Times* (Moscow), February 1976, p. 1.
11. See Schlesinger press conferences of 30 November 1973, 10 and 24 January 1974, as reported in the *New York Times*' next-day issues; or see report in *Survival*, March–April 1974.
12. See "Soviet Attitudes to Controlled Strategic Conflict," *Current Comment*, Carleton University School of International Affairs (Ottawa), May 1976.
13. M. A. Milshteyn and L. S. Semeyko, "The Problem of the Inadmissibility of a Nuclear Conflict," translated from the *IUSA* (Moscow) *Journal*, in *Strategic Review*, Spring 1975.
14. See discussion in this author's "The US-Soviet Strategic Balance," *Current History*, October 1980.
15. The British, American, and Canadian "Sovietologists" assembled by The University of New Brunswick's Centre for Conflict Studies (Fredericton) in March 1980, for a specialist seminar discussing appropriate Western low-intensity strategies, were unanimous on this point; see published Conference Report, *Low Intensity Conflict and the Integrity of the Soviet Bloc*, UNB, 1981.
16. The 1970s editorials and analyses of the noted anti-Soviet Marxist journal *Critique*, of Glasgow, Scotland, were illustrative.
17. See data in this author's "Japanese Security in a Changing World: The Crucible of the Washington-Moscow-Peking 'Triangle' ", *Pacific Community*, April 1975.
18. Note the flow of detailed reports on East–West trade prospects and problems from Carleton University's Institute for Soviet and East European Studies (under the general editorship of Dr. C. McMillan, Director of the Institute). See also, e.g., *Soviet Chemical Equipment Purchases From the West: Impact on Production and Foreign Trade,* CIA, Langley, Va., October 1978; and Philip Hanson, "Western Technology in the Soviet Economy," *Problems of Communism*, November–December 1978, pp. 20–30.
19. "Science Brief—Russia's Research Harvest," *The Economist*, 30 May 1981.
20. The Soviet interventions in Angola and Ethiopia are analyzed in depth in "Soviet

Strategic Initiatives," Chapter 8 ("Moscow and the Southern Flank: New Posture, New Capabilities").

21. The one-sidedness of the dependency relationship, from the mid-1960s through the mid-1970s and into the late 1970s, had allowed Moscow to curb Castro's impetuous (and independent) pursuit of the larger revolutionary vistas conjured up by Latin American sociopolitical realities. The late 1970s and early 1980s saw the development of a new pattern. In South America, Cuba continued to act with circumspection in deference to Moscow's pursuit of accommodation with at least some elements of the traditional elites, and improved trade and other state-to-state relationships (as in the cases of Peru and Argentina), but Central America witnessed renewed Cuban activism. It was not a return to the early 1960s phenomenon of competitive Soviet and Cuban pursuit of distinct and sometimes contradictory interests. The new phenomenon was rather one of clearly delineated spheres of activities and interests, a testimony to Soviet–Cuban negotiation, and a more balanced dependency relationship.
22. See Jacobsen, "The US-Soviet Strategic Balance."
23. For a thorough analysis, see Donald M. Snow, "The MX-Basing Mode Muddle, Issues and Alternatives," *Current News Special Edition,* US Department of Defense, Washington, D.C., 2 October 1980; for a somewhat different perspective, see Snow's "MX: Maginot Line of the 1980s," *The Bulletin of the Atomic Scientists,* November 1980, pp. 20–25, and the 20 August 1981 report issued by the Interior Committee of the US House of Representatives (which stated that existing plans were "strategically defective" and would create only "an illusion of survivability").
24. See, for example, the special feature on "Reagan's Defense Buildup" in *Newsweek*, 8 June 1981.
25. See, e.g., E. Ulsamer, "In Focus," *Air Force Magazine*, November 1980.
26. Dr. W. J. Perry's statement in US Senate, *Department of Defense Appropriations Fiscal Year 1980* (96th Congress, First Session), Part 4, p. 43.
27. By 1980 then-existing nuclear arsenals translated into more than 30 tons of TNT for every man, woman, and child on earth (estimate made by Paul Warnke, former Director of the US Government Arms Control and Disarmament Agency, public lecture, Miami, 17 September 1981). See H. J. Geiger, "The Illusion of Survival" and M. Shuman, "The Mouse that Roared," in *The Bulletin of the Atomic Scientists*, June/July 1981, pp. 16–19 and January 1981, pp. 15–21, respectively. Also see K. N. Lewis, "The Prompt and Delayed Effects of Nuclear War," *Scientific American*, July 1979.
28. Note Andrew Cockburn and Alexander Cockburn, "The Myth of Missile Accuracy," *The New York Review of Books*, 20 November 1980, pp. 40–43.
29. See, e.g., *International Security*, Winter 1979/80; articles by Lord Mountbatten and Richard Garwin.
30. Text of acceptance speech by George Kennan on receiving the Albert Einstein Peace Prize in Washington, 19 May 1981; for full transcript, see, e.g., the (British) *Guardian*, 25 May 1981.
31. *Estimated Soviet Defense Spending in Rubles, 1970–75,* CIA, Langley, Va., May 1976; and see A. M. Cox, "Why the US Since 1917 Has Been Misperceiving Soviet Military Strength," *New York Times*, 20 October 1980.
32. One should note that even if one were to accept the B-team's figures for Soviet defense spending, total NATO expenditures still would exceed total Warsaw Pact expenditures by a considerable margin (as they have every year since 1970!). See table, "NATO Outspends Warsaw," *The Bulletin of the Atomic Scientists*, June/July 1981; also see *World Military Expenditures and Arms Transfers 1969–1978*, US

Arms Control and Disarmament Agency, Washington, D.C., 1979, and former US Secretary of Defense H. Brown's final *Report* to Congress, January 1981.

33. See especially F. D. Holzman's "Are the Soviets Really Outspending the US on Defense?" *International Security*, Spring 1980; and J. O'Grady's "The CIA and the Soviet Military Budget: Who's Fooling Who?" unpublished CDA analysis, Summer 1981. As concerns the CIA's politically most potent comparisons, namely those based on "dollar costing," the 1981 SIPRI Yearbook sums up the scientific consensus: ". . . This is a wholly illegitimate method of international comparison" (Stockholm International Peace Research Institute, *World Armaments and Disarmament: SIPRI Yearbook 1981*, Taylor & Francis, London, and Oelgeschlager, Gunn & Hain, Cambridge, Mass., 1981).
34. CIA report, September 1980; see discussion in the (Toronto) *Globe and Mail*, 25 March 1981; and note "Soviet Energy Options and United States Interests," *Quarterly Report*, Mershon Center, Ohio State University, Summer 1980. For original (pessimistic) analysis, see "Prospects for Soviet Oil Production, A Supplementary Analysis," CIA report ER77-10425, July 1977.
35. The annual *Jane's Fighting Ships*, London, provides data on Naval ship numbers and characteristics. See also SIPRI Yearbooks.
36. C. G. Jacobsen, *Soviet Strategic Initiatives*, especially Chapters 3, 4, and 8.
37. "Soviet Aid Disputed in Terrorism Study," *New York Times*, 29 March 1981.
38. See, for example, R. B. Laird's "The All-Volunteer Force: It Isn't Working," *International Security Review*, Spring 1981; also Barry M. Blechman and Leslie H. Gelb, "Our Weakened Defenses—Money Alone Is Not the Answer," *Forbes*, 13 October 1980.
39. *Newsweek*, 8 June 1981, p. 29.
40. "Soviet Armed Forces Showing Weaknesses in Several Key Areas," *New York Times*, 9 December 1980; and note Secretary of Defense H. Brown's testimony, as reported in the *Chicago Tribune*, 14 October 1980. For more detail, see transcript of *World Special: The Red Army*, PBS (Public Broadcasting System), Boston, 1981.
41. See, for example, John Parmentola and Kosta Tsipis, "Particle Beam Weapons," *Scientific American*, April 1979; Richard Burt, "Experts Believe Laser Weapons Could Transform Warfare in 80's," *New York Times*, 10 February 1980; and contrast with "Beam Weapons Technology Expanding" section, *Aviation Week and Space Technology*, 25 May 1981.
42. Richard L. Garwin, "Are We on the Verge of a New Arms Race in Space?" Special Report, *The Bulletin of the Atomic Scientists*, May 1981.
43. Frank Barnaby, "World Arsenals in 1981," *The Bulletin of the Atomic Scientists*, August/September 1981, pp. 16–20; note also Ruth Leger Sivard's *World Military and Social Expenditures 1981*, WMSE Publications, Leesburg, Va., and the 1981 SIPRI Yearbook.
44. "Rethinking the Unthinkable," *New York Times Magazine*, 15 March 1981, p. 70.
45. See table, "US-Soviet Modernization of European Missiles," *The Bulletin of the Atomic Scientists*, October 1980, p. 27. One might add that in one sense SS-20 mobility was a "threat" only if NATO planned to initiate hostilities. A NATO first strike could have taken out older, immobile Soviet "Euromissiles"; the mobile SS-20s might survive. If one feared a Soviet first strike, however, the SS-20s might be preferable. Their greater accuracy dispensed with the need to compensate for poor targeting by increasing warhead yields. They promised to entail less "collateral" damage than their predecessors.
46. See, for example, S. Meikle, "Has Marxism a Future?" *Critique* (Glasgow), No. 13, Spring 1981, pp. 103–122; and "Brazil Church's Voice of Dissent Growing," in

Around the Americas, *Miami Herald*, 19 April 1981.

47. E. P. Thompson, "The END of the Line," *The Bulletin of Atomic Scientists*, January 1981; and "Neutralism: The Task of Remaking Europe," *The Guardian* (London and Manchester), 6 September 1981.
48. The *New York Times,* 10 May 1982, provided a transcript of the President's address; *CNN* (Cable News Network), 9 May 1982, aired Muskie's response; The *New York Times,* 12 May 1982, reported Haig's follow-up.

Chapter 4

Other Nuclear Powers: China, Britain, France, India, Israel(?), The Republic of South Africa(?) —Capabilities and Restraints

Ever since its first nuclear explosion in 1964 (in one of history's more startling ironies, the same day that Khrushchev, Mao's old adversary, was ousted), China has loomed as the world's "third" nuclear power. But while China's ultimate superpower potential could not be denied, the military reality of its early 1980s capabilities placed it squarely within the ranks of second-echelon powers. China's more august status owed more to psychological and other considerations than it did to military prowess.

In analyzing the "second-echelon" powers, it would seem appropriate to deal with them in chronological order, according to the date of their entry into the "nuclear club."

British (and Canadian) scientists were central to the Allied bomb development effort during World War II. The program was moved to the United States (where it became known as the Manhattan Project) because British facilities were considered to be too exposed to possible Luftwaffe attack. The McMahon Energy Act passed by the US Congress after the war's conclusion, which enjoined the American government not to share nuclear design information with foe *or* ally, was therefore seen as preposterous ("infernal cheek") by London. The estrangement from Washington quickly waned, however, under the impact of postwar reconstruction demands, the enfeebling of old empire bonds and power, and the darkening shadow of cold war polarization.

Britain's first atomic bomb was detonated in October 1952, and its first hydrogen bomb four years later.[1] The earliest bombs were carried on American B-29s, which had been transferred to RAF Bomber Command in 1950. By 1955 the swept-wing (British) Valiant jet bombers were entering service, followed two years later by the delta-wing Vulcans, and in 1958 by the Victors. 1962 saw a newer version of the Vulcan, no longer restricted to free-fall weapons delivery, but capable (as were the Victors) of employing air-to-surface missiles. The early 1960s brought Bomber Command strength to its peak of 180 Vulcan and Victor planes, many armed with "stand-off" Blue Steel missiles with a range of 200 nautical miles. But penetration of rapidly improving Soviet air defense capabilities soon came to be viewed as an ever more dubious proposition. For a while it was thought that the American Skybolt air-to-ground missile, with its projected range of 1000 nautical miles, might keep bomber prospects alive. The Kennedy administration, however, aborted the program.

In the meantime Britain had pursued development of an extraordinarily advanced TSR-2 strategic bomber capable of Mach 2 speeds, with a combat radius of more than 1000 miles, and the Blue Streak intermediate-range missile, designed to be fired from underground silos. But the increased skepticism concerning the future viability of bombers, however sophisticated, led inexorably to the cancellation of the TSR-2. And soon the Blue Streak was also cancelled. It was victim to the persuasion that improving Soviet accuracy potentials would threaten even silo-based missiles, and that possible Soviet launch sites were too close to allow for adequate warning.

Consequently, in 1962 Britain made the decision to switch to a submarine deterrent. Polaris nuclear-class submarines and submarine-launched ballistic missiles were acquired with American help; the warheads were produced by Britain. The first vessel was completed in 1968. The program was to have encompassed five submarines, allowing two to be at sea at all times. But 1965 brought a decision to limit production to four (the last was completed in 1969). This meant that only one submarine could be guaranteed to be at sea, though it would intermittently be joined by a second vessel.

In 1980 the UK government decided to replace aging Polaris boats with the more formidable Trident submarines, armed with multiple independently targetable reentry vehicles (MIRVs); the Tridents were to be in place by the early 1990s. Again, original aspirations called for five vessels. But financial considerations were seen to ordain a maximum of four, thus ensuring that "on-station" deployment prospects would remain limited to "one and a half." The cost, projected at £4 to £5 billion in July 1980, was within a year calculated more realistically at between £8 billion and £10.785 billion (between about $16 billion and $22.5 billion).[2]

Before further comments on prospects and restraints, one should perhaps revert attention to the early 1960s. In the 1960s, as the Blue Streak was being cancelled, Britain accepted deployment on its soil of American Thor missiles; they remained until 1963. But the Blue Streak concellation, though encouraged by cost considerations, was, as mentioned, motivated primarily by the judgment that land deployment this close to Soviet firing locales did not make military sense. One is therefore forced to conclude that the Thor deployment was essentially a political exercise, nurtured by the political policy requirements of the American–British relationship and the NATO context. (The Thor phenomenon found its analogy in the early 1980s refusal to disband America's land-based missile forces. Their greater distance from Soviet launching platforms had once made them more viable, yet now-developing Soviet potentials were threatening them also—at least theoretically. The major reason for their survival in an era favoring sea and air potentials was clearly rooted in the political sanctity of the old "Triad" concept, in interservice and bureaucratic politics.)

A similar comment must be made concerning the Thatcher government's 1979 cruise missile decision; the government committed itself to accept American cruise missile deployment in 1983. Twenty-some years after military considerations were seen to have dictated a move away from land deployment in Britain, such a venture was again ordained. The cruise missiles were of course mobile, and hence not as easily targetable as silo-based missiles. But the realities of British society and British geography circumscribed the military authority's ability to draw optimum advantage from this design characteristic. Considering Soviet data-gathering capabilities, and the built-in redundancy of Soviet force potentials, there were many who thought that cruise mobility would not preclude hostile targeting; rather, it might merely ensure that attack designs would include more and larger bombs or warheads, so as to preclude the possibility of any slipping out of the net.[3] Cruise deployment could well be argued to establish a new "need" for Moscow to target British land, a need that had not previously been manifest, and furthermore to guarantee that "collateral" destruction would be extensive. The Labor Party's opposition to the scheme was not unreasonable, not if military rationales were the sole criteria. The fact that the government nevertheless chose this path was, again, a reflection of the preeminence of politics. Land-based cruise missiles could perform no military mission that could not be performed (better) by submarine-launched ballistic missiles, or by sea- or air-based cruise versions.

In fact, even the core of Britain's deterrence, the submarine force, was somewhat suspect. The perhaps three vessels in port (or repair yard) would be vulnerable. The fourth could be subject to sufficient concentrated surveillance to give the advantage to hostile anti-submarine warfare forces. Furthermore, the single warheads of the limited number of missiles aboard the Polaris boats might well not have sufficed to penetrate the rudimentary missile defense system guarding Moscow. MIRV replacements embodied more secure penetration prospects (presuming successful launch), but the comprehensive and ambitious nature of ongoing Soviet ballistic missile defense research constituted a potential caveat of some import.

British atomic potentials clearly entailed a probably decisive power differential vis-à-vis non-nuclear powers. But when compared to superpower capabilities, they were as a (single) Lilliputian up against giant Gullivers.[4]

Indeed, by 1981, even Establishment sources spoke more of the prestige associated with purportedly independent nuclear forces than of their military relevance. Trident costs were justified by the program's presumed impact on national prestige; the program was not so much a deterrent as a status symbol.[5]

The French effort, across the channel, was analogous yet distinct. France's claim to independent deterrent status was more serious. France exploded its first nuclear device in February 1960. The decision was a function of DeGaulle's conviction that American security guarantees were no longer reliable. The conviction drew sustenance partially from World War II memories and partially from independent analyses of the nature and impact of the evolving US–Soviet "balance"; it was to be reinforced by Washington's refusal to share nuclear-related information with Paris.

The former French Chief of Staff, Pierre M. Gallois, has presented a succinct summary of the strategic analysis:

> As long as the territory of the United States was outside the reach of Soviet missiles . . . it was clear that the alliance sealed by Washington was militarily totally effective. . . . A situation existed of total strategic asymmetry, which was favorable to the defense of the West.
>
> In the beginning of the 1960s . . . the United States lost its geographical invulnerability. . . . Having become vulnerable in its own territory, the United States could no longer provide European allies with unconditional guarantees as to their defense.

> . . . [American] strategy moved on to "flexible response." . . . Europeans, the French leaders in particular, saw in this change . . . confirmation of the fears they had expressed some years before.
>
> It was necessary that, in extreme circumstances, by brandishing the threat of an avenging spasm, France should be capable of inspiring a certain amount of fear.
>
> If the new vulnerability of the United States and the dwindling of the guarantees offered to its allies led . . . [to a French] national nuclear force, the blunders of American foreign policy also contributed. [Washington's and London's 1958 intervention in Lebanon was staged] without warning Paris. . . . One month later, at the time of the bombardment of the islands of Quemoy and Matsu, the American government brandished nuclear warheads, naturally without consulting, nor even informing, French allies.[6]

Suspicion that American security guarantees were hollow and that the United States would not put Washington at risk for the sake of Paris combined with fear that France might find itself under nuclear bombardment as a consequence of Allied actions over which it had no control, and more general chauvinist resentment at perceived high-handedness. De Gaulle flamboyantly asserted French independence by ordering the indigenous nuclear arsenal, by declaring that France's targeting policy would be one of *"tous azimuths"* (all-directional, ready to meet hypothetical threats from West as well as East), and, ultimately, by withdrawing from NATO's military structure. Unofficially, there was scant doubt even under De Gaulle that the primary "threat" was still thought to come from the East. De Gaulle continued to sanction a degree of cooperation with NATO's military authority, and the cooperation was extended by his successors. Nevertheless, his *tous azimuths* proclamation underlined French alienation and enthnocentrism, and French determination to seek self-reliance.

Early French bomb capacities were deployed aboard Mirage bombers, and the air element was to remain central to France's strategic posture. At the dawn of the eighties France had 33 Mirage IVA strategic bombers, harboring two-thirds of its nuclear arsenal.

Deployment of silo-based SSBS (Sol-Sol Balistique Stratégique) missiles, with 1500-mile range and 150-kiloton warheads, began in 1971, on the Plateau d'Albion. Obviously, the French were less concerned about emerging Soviet accuracy technologies than the British (and their apparent sang-froid was perhaps justified; see Chapter 3's discussion of accuracy measurements, and differences between wartime and peacetime test trajectories). Still, a certain unease about their survivability was evident in the fact that actual deployment numbers were to lag well behind original projections. Early plans called for 27 missiles, 18 of which were in place by 1974. The final 9 were in fact never deployed; the figure remained 18 through the remainder of the decade.

1970 saw the first French (nuclear) strategic missile submarine, Le Redoutable. Five were projected by the decade's end, and five were procured. Thus, as opposed to the British, the French could rely on two being at sea at all times, to be joined occasionally by a third.

France's nuclear program continued to receive far more extensive funding than the British. By the end of the 1970s the French were pursuing multiple independently targetable warhead technologies, new bomb variants (including the "neutron" bomb), the possibility of additional submarines, and continued extensive testing. In 1979, for example, France authorized nine test explosions; the UK, one.[7]

Nevertheless, one could argue that "the 18 land-based intermediate-range ballistic missiles . . . and the 33 Mirage IVA aircraft . . . are more targets than a deterrent," submarines in the pen "are subject to almost instant destruction" and "those at sea can be tracked by Soviet killer subs and must operate in narrow and detectable sea channels." Further, "Gallois has himself admitted the vulnerability of French submarine forces and has advocated large increases in their number and capability. (See, for example, the May 1980 issue of the French opinion journal *Paradoxes*)."[8]

There is one peculiarity to early French nuclear policy that deserves separate mention, namely the "trip-wire" concept. This held that aggression would be met by immediate nuclear response. It reflected French suspicion that America's embrace of "flexible response," the move away from Eisenhower's "Massive Retaliation" doctrine, uncoupled Europe from the US–Soviet context, allowing for localized European war. The trip-wire concept derived from the expectation that nuclear exchanges would inevitably escalate to the strategic level. Paris was in effect saying, "We'll do it if you don't; war in Europe will not be contained." The French stance was emasculated, however, by subsequent developments. Deciding that conventional force levels would remain less than envisaged by proponents of flexible response, NATO soon announced that the Western alliance would in fact itself initiate the use of nuclear weapons to stem "aggression." France's trip-wire policy thus took on an air of redundancy. But the concern that had occasioned the policy remained alive. NATO's "first-use" dictum was accompanied by stress on "battlefield" and "theater-nuclear weapons." NATO was, in other words, espousing as official policy the view that escalation from "theater-nuclear exchanges" to "strategic exchange" need not be automatic, that even nuclear war in Europe could be contained. Independent observers greeted the thesis with extreme skepticism (see Chapter 5). Yet the dubiousness of the premise, or postulate, merely reinforced France's cynicism about America's willingness to live up to alliance commitments. Hence the advocacies for a further expansion of indigenous capabilities, to shore up their credibility.

But the ideals of true strategic independence foundered on geographical and socioeconomic realities and constraints. President Valéry Giscard-d'Estaing's late 1970s policy of ever-closer cooperation with NATO military councils reflected official recognition of French limitations. And his successor, François Mitterand, was clearly even less disposed to pursue De Gaulle's grander vision. France's posture might have gained somewhat greater credibility than Britain's, but the difference was one of relativity, not essence; its forces' survival prospects remained open to doubt, its penetration prospects remained far from secure.

China's position was similar. Its first explosion, in 1964, had been met by considerably greater awe than greeted British or French efforts. The image of the old Middle Kingdom, its immense population, its vast territories and resource potential, and its obvious independence from Western and Eastern "bloc" constellations conjured up a superpower aura. It was perhaps inevitable that Chinese prospects would be exaggerated. And they were. The late 1960s, for example, brought US Defense Department analyses projecting significant Chinese intercontinental-range ballistic missile (ICBM) numbers within 10 years (estimates ranged from 10 to 75).[9] Yet it was to be 1980 before China was able to deploy its first two, perhaps three, primitive first-generational ICBMs. Early 1980s evidence indicated that further increases would be less than impressive; significant numbers were not likely to be attained before the 1990s, if by then.

As of the early 1980s China still relied on older-technology (Soviet-model and Soviet-derived) bombers that lacked modern stand-off capacity. Their penetration prospect against northern defense deployments was highly questionable. The scope of Soviet warning systems (satellites, radars sweeping into China from near-border locations, and regularized overflights over Manchuria, Inner Mongolia, and Sinkiang) promised to allow more than sufficient time for the activation of sophisticated fighter and missile defense means.

China's early-generational, medium- and (few) intermediate-range missiles were hardly more impressive. They were faulty, encumbered by "degradation factors" that ensured that only a limited percentage would reach designated targets in the best of circumstances; their fuel technologies were equally "developmental," a fact that inevitably undermined their potential state of readiness; and finally, their command and control facilities did not yet incorporate the sophisticated technologies that allowed the superpowers to envisage minimal activation times (see Chapter 2). Chinese firing preparation time requirements were such that one could conceive of Soviet preempt before the process was complete. The potential speed of Soviet intelligence, Soviet data processing, analysis, and decision, and Soviet missile activation sufficed to make possible Chinese conflict initiation a high-risk proposition. China attempted to circumvent technological limitations through innovative dispersal and camouflage techniques, in craggy mountainous terrain. But its hope was restricted to that of increasing Soviet uncertainty and hence denting Soviet confidence; its potential remained less than assured.

China's attempt to procure a more survivable submarine-based missile force stumbled. It still had one old G-class submarine with missile-launching tubes acquired from the Soviet Union, but Moscow had not provided missiles, or suitable warheads, and attempts to build these repeatedly failed. The (until then) most ambitious effort apparently ended when underwater firing tests resulted in the explosion of the vessel and death of the approximately 100-man (woman?) crew in September 1981.[10]

Chinese prospects were also somewhat stymied by another factor, Soviet ballistic missile defenses. As previously indicated, the limited ballistic missile defense system around Moscow had proven potency against single-warhead missiles. And China remained (very) far from mastering the MIRV technologies that had emasculated the system's potency vis-à-vis American penetration means. In September 1981 the Chinese finally succeeded in launching three satellites with one booster rocket. But this merely proved that they could deploy free-fall multiple re-entry vehicles aboard their missiles, a far cry from MIRVs; existing Soviet defense prospects were not significantly affected (see Chapter 3).

The ICBMs deployed by China during the 1980s were likely to remain of the single-warhead or MRV variety. And they were unlikely to exceed the figure needed to saturate Soviet anti-ballistic missile numbers, even if these remained within the limit of 100 sanctioned by the now-defunct SALT I Treaty (as amended in 1974). There was the point, moreover, that even Chinese success in procuring greater numbers and, one day, perhaps even (limited) MIRV potentials, might not assure penetration. Moscow had throughout the 1970s poured considerable funding into BMD research, testing, and development, especially at the Sary Shagan range in Central Asia (conveniently located under probable Chinese flight trajectory paths). By 1980–81, amid intensified Soviet (and now also American) BMD effort and interest, there were reports that Moscow had tested newer-technology mobile ABMs and much-improved radars, along with indications of progress toward the development of revolutionary laser and high-energy particle beam defense concepts.

Chinese capabilities in 1980–81 were remarkably similar to those of the USSR in 1957. Yet even this analogy was flattering, since China's technological-economic infrastructure remained less potent than Moscow's had been at that time. Chinese logistics capabilities were, for example, inferior to Soviet standards of 1945. The 1957 analogy is doubly flattering if taken as an indicator of China's ability to close the gap. Moscow had enjoyed three important advantages in 1957. Its adversary at the time did not yet possess the accuracy technologies that would later add doubts to land missile survivability prospects; nor did its adversary have even rudimentary missile defenses; nor did it have the sophisticated satellite means that would bring intelligence confidence. China's northern adversary two and a half decades later did.

There were those who felt that China's improving access to Western technology might bring significant early results. Yet the evidence indicated that China's military-industrial absorption capacities (as also those of its civilian sector) were limited. The military impact of China's leery rapprochement with the West was likely to be relative, and incremental at best; it was not likely to bring about a dramatic short-term, qualitative change.[11]

China's potential, like that of the more sophisticated systems available to Britain and France, thus remained marginal and uncertain. Strike prospects against Soviet targets were limited mainly to Soviet provincial regions and cities; chances of success against Moscow itself remained dubious. All three nations' deterrence rested, at best, on the ability to inject a limited measure of uncertainty into Soviet planning. And China appeared to recognize that this was the most that could be hoped for, for the foreseeable future. 1980–81 brought a near 20-percent *decrease* in China's military budget.[12] Logic decreed the conclusion that its fear of Soviet aggression had lessened; its military ambition had perforce become more modest; civilian economic aspirations demanded greater priority. Of the three medium nuclear powers, China's ultimate potential was the only one that encompassed true superpower possibilities. But the realization of such ambition was still only a gleam in the eye, a matter of possible twenty-first-century impact.

India was in a different class, though its potential was, perhaps, similar. The first Indian nuclear device was detonated in 1974. It was described as a "peaceful nuclear explosion," intended to harness nuclear energy for the purpose of civilian developmental ventures (canal dredging and whatnot). And India made no immediate attempt to proceed to the stage of force deployment. Nevertheless, the explosion's psychological impact abroad was obvious, and surely intended. It served to fuel nuclear aspirations in both Iran and Pakistan. By 1981–82 when it became clear that a Pakistani bomb-development effort was on the verge of success, Delhi began to move toward weapon systems deployment.

In some ways India's nuclear potential had long exceeded China's.[13] India's nuclear program dated back to 1944. The Indians could probably have exploded a device as early as 1960, before France, and well before China; that they did not do so was due to political considerations, not technological limitations. The emphasis throughout was on civilian utility, nuclear power for energy generation (China's civilian reactor program was still in its infancy, still in the early developmental stage, at the dawn of the eighties), and missile and satellite technology for education and research. The weapons potential was a "by-product." Partly because it could be argued to have come gratis and partly because nuclear arms ambitions were disavowed, the public opposition that might otherwise have been generated in a society infused with the legacy of Mahatma Gandhi never developed. Government spokesmen only began to discuss weapons deployment after Pakistani developmental efforts became manifest; in the new context, the legacy of past wars was perhaps more viscerally potent than that of past ideals.

Two other states are generally presumed to have acquired limited nuclear arsenals during the 1970s, Israel and South Africa. The original Israeli bomb development is thought by the CIA and other observers to have been made possible by the hijacking of the *Shersburgh*, a ship carrying 200 tons of plutonium from West Germany for Euratom, and possibly by the acquisition of yet more plutonium from other clandestine sources and operations.[14] Since the 1964 activation of the French-built Dimona research reactor, Israel has also had a steady source of domestic plutonium. It has been estimated that Dimona produces sufficient plutonium for one bomb every year.[15] By 1975 evidence emerged that Israel might have developed a capacity for uranium enrichment.[16] The following year the CIA estimated that Israel had 10 to 20 nuclear weapons.[17] The Israeli government denied having proceeded from weapons potential to weapons production, but the weight of testimony was one-sided, to the contrary, and compelling. Until 1979 there was no evidence of an Israeli bomb test, though the quality of the Israeli scientific establishment was thought to ensure confidence in the product. Still, cautious defense planners would obviously desire physical evidence, if politically possible.

The presumed search for testing data ties in both with South African developments and with Israel's emergence as the Republic of South Africa's most intimate military supplier and collaborator. On August 6, 1977 . . . "the Soviet Union informed the United States . . . that Soviet intelligence had spotted installations for detonating an explosion in the Kalahari desert. The United States quickly redirected satellite cameras and verified the information"; Washington then "coordinated . . . a strong joint protest to South Africa not to proceed."[18] Pretoria denied that an explosion had been intended; none took place, then. It should be noted, however, that Pretoria never offered any alternative explanation for the structures revealed by Soviet and American satellite photography.

Two years later came evidence of a possible nuclear test (or tests) over southern ocean waters. "The intense burst of light was picked up on September 22 by a Vela satellite, which is specifically designed to pick up nuclear tests. It was scanning a 4500 square mile swathe below, most of it sea, but including southern Africa. . . . The pulsating flash—it has also been described as two flashes—strongly suggested a nuclear explosion."[19] "US experts . . . concluded that the 1979 event was a neutron bomb test . . . conducted jointly with Israel."[20] Early US government reports to that effect were denied by Tel Aviv, as well as by Pretoria, and there followed a period of conflicting "interpretations." Having first asserted that the evidence was conclusive, the Defense Intelligence Agency (DIA) backtracked, saying it did not constitute definitive proof. The CIA followed up with a statement that available evidence was insufficient. But then the DIA reverted to its original stand, that the evidence was indeed sufficient, and convincing. There the matter stood.

NOTES

1. See Norman Polmar, *Strategic Weapons: An Introduction,* Crane, Russak & Company, New York, 1975, pp. 67–72.
2. The initial, government estimate is taken from "It's Trident, It's American, and It's a Bargain," *The Economist,* 19 July 1980; the lower "revision" was prepared by David Greenwood, Director of Aberdeen University's Centre for Defence Studies (see *The Guardian,* September 20, 1981), while the higher estimate was calculated by Judith Reppy and Harry Dean, of the Armament and Disarmament Information Unit in the Science Policy Research center at Sussex University (see note 3).
3. In fact, recognition of this proposition was integral to the government's decision to focus British funds on modernizing the submarine-based forces, rather than on cruise missile alternatives; see, e.g., J. Reppy and H. Dean, "Britain Buys the Trident," *The Bulletin of the Atomic Scientists,* November 1980, pp. 26–31.
4. American attitudes are revealed in the fact that they tended to class Britain's capabilities as part of NATO's "theatre-nuclear forces," rather than as "strategic" force elements. Ibid., p. 27.

5. D. Dilks, *Retreat from Power: Studies in Britain's Foreign Policy of the Twentieth Century,* Macmillan, London; see review in *The Economist,* 28 February 1981.
6. Pierre M. Gallois, "French Military Politics," *The Bulletin of the Atomic Scientists,* August/September 1981, pp. 21–25.
7. Frank Barnaby, "World Arsenals in 1980," *The Bulletin of the Atomic Scientists,* September 1981, p. 14.
8. Edward A. Kolodziej, "Furthermore . . ." (riposte to Gallois's "French Military Politics"), *The Bulletin of the Atomic Scientists,* August/September 1981, p. 21. The combination of skepticism about US resolve and doubt as to the efficacy of indigenous French means was reflected in France's cautious yet ambivalent early-1980s support for American cruise and Pershing II missile deployment plans. On the one hand Paris saw these plans as still another indication of Washington's desire to decouple European war prospects from US home security. On the other hand, the fact that the missiles would be targeted on Soviet locales suggested immediate Soviet retaliation against US targets. France's stand echoed Cuba's posture of 1962.
9. See Pentagon charts in C. G. Jacobsen, *Soviet Strategy–Soviet Foreign Policy,* Robert MacLehose, The University Press, Glasgow, 2nd ed., 1974, p. 108.
10. ABC *World News,* 15 October 1981, quoting Japanese intelligence sources.
11. See C. G. Jacobsen, *Sino-Soviet Relations Since Mao: The Chairman's Legacy,* Praeger Publishers, New York, 1981, especially Chapter 1, "The Strategic Context."
12. Deputy Prime Minister Yao Yilin, speech to the Standing Committee of the National People's Congress, 28 February 1981; see, e.g., report in the *New York Times,* 1 March 1981. See also O. Edmund Chubb, "America's China Policy," *Current History,* September 1981, p. 281: "In September [1980] . . . the National People's Congress decided to cut the military budget by $2 billion, reducing it to $13.1 billion."
13. Onkar Marwah, "India's Nuclear Program: Decisions, Intent and Policy, 1950–1976," in W. H. Overholt, ed., *Asia's Nuclear Future,* Westview Press, Boulder, Colo., 1977, pp. 161–196.
14. See George H. Quester, "Israel and the Nuclear Non-Proliferation Treaty," *The Bulletin of the Atomic Scientists,* June 1969, p. 9; Shlomo Aronson, *Israel's Nuclear Option,* Center for Arms Control and International Security, UCLA, November 1977; and "Interview," *The Search, Journal of Arab and Islamic Studies,* April 1981.
15. Shai Feldman, "A Nuclear Middle East," *Survival,* May/June 1981.
16. R. J. Prangler and D. R. Tahtinen, *Nuclear Threat in the Middle East,* American Enterprise Institute for Public Policy Research, Washington, D.C., 1975.
17. A. Kranish, "CIA Sees 10–20 Israeli A-Arms," *Los Angeles Times,* 15 March 1976. The first, only slightly less authoritative report that Israel possessed nuclear bombs had appeared nearly a year earlier, in August 1975, when the *Boston Globe* reported that Israel had an arsenal of 10 nuclear weapons; the author was William Beecher, former *New York Times* correspondent and Deputy Assistant Secretary of Defense (from 1973 to May 1975), who had recently returned from a visit to Israel and Egypt. See Polmar, *Strategic Weapons,* p. 78. US government conviction that Israel possessed nuclear arms apparently dated back to 1974; see "3 Nations Widening Nuclear Contacts," *New York Times,* 28 June 1981.
18. *South Africa: Time Running Out,* The Report of the Study Commission on US Policy Toward Southern Africa, Foreign Policy Study Foundation, Inc., University of California Press, Berkeley, 1981, p. 358; see also "3 Nations Widening Nuclear Contacts."
19. *The Economist,* 3–9 November 1979.
20. *South Africa: Time Running Out,* p. 252; also "3 Nations Widening Nuclear Contacts."

Chapter 5

Nuclear Proliferation—Past the Point of No Return?

By the late 1970s nuclear scientists attending the Pugwash symposia and other world gatherings were warning that nuclear war before the end of the century had become a probability, not just a possibility. Scenarios tended to postulate (initially) limited nuclear conflicts in distant arenas, rather than an immediate superpower Armageddon. But there was a consensus of dread, summed up by Britain's International Institute of Strategic Studies in 1981, that "even a modest exchange of nuclear weapons would, in all probability, escalate rapidly to the strategic nuclear level."[1]

In a special issue of the *Annals* of the American Academy of Sciences[2] and in other scientific journals and fora, specialists discussed the likely number of nuclear weapon states toward the end of the century. Projections tended to coalesce at 20 to 30, but ranged up to 50.

The scientific knowledge needed to build nuclear bombs had long been available: "As far back as 1967 Sir John Cockroft, Britain's top nuclear research executive, admitted 'there are no secrets' to prevent any nation from developing bombs."[3] The point was hammered home during the 1970s, when a number of students from Princeton and other universities proved able to construct realistic bomb designs from information available in the open stacks of the Library of Congress. Journalists proved equally able to ferret out the "secret."[4]

In the mid 1970s the Stockholm International Peace Research Institute calculated that by 1980, 30 nations would, theoretically, be able to build small nuclear arsenals.[5] By 1980 an analysis in *Scientific American* documented that 40 countries could, if they wished, procure nuclear weapons within 10 years.[6]

Most nations that seemed within grasp of nuclear status had until the early 1980s been deterred by one of three considerations. Some, like Norway, Sweden, Denmark, Holland, Canada, and Japan, desisted from pursuing nuclear potentials primarily for political reasons, and because of moral abhorrence. Some desisted because they did not have and could not secure access to plutonium or alternate nuclear materials. Finally, most realized that while nuclear bombs might somehow be built or acquired, it was little good having bombs if one did not also have credible means of delivering them to relevant targets; sophisticated delivery systems, whether based on missiles or penetration bombers, had generally been beyond reach.

By the late 1970s, however, the latter constraints were crumbling, if not already shattered, and the consequence of this state of affairs promised also to gnaw away at self-imposed political and moral restraints. Journalists from the *New York Times* and other newspapers had proved able to walk and bluff their way into American nuclear bomb storage depots in Western Europe. Security at such storage depots remained suspect. There was all too much reason to believe that bombs could one day be stolen, by terrorist or other groups with allegiance to nations, causes, or individuals, to ideals, or to avarice.[7]

But the bomb was also becoming easier to build. The technical data had, as mentioned, long been available. Now, enriched uranium and plutonium were also becoming more common. Israel's presumed nuclear stockpile was thought by the CIA to owe its existence to the hijacking of a shipload of plutonium.[8] Apart from that particular shipload, considerable amounts of plutonium were "missing" from civilian nuclear reactor fuel waste storage facilities, more than could possibly be accounted for by even the most generous estimates of "natural" disappearance.[9] Other plutonium transport ventures, by ship and by rail, had likewise resulted in "losses" of greater or smaller amounts en route. Finally, the proliferation of "civilian" nuclear reactor capabilities in developing nations was, of course, also resulting in a proliferation of plutonium "waste" and "side-product" depots in more and more countries; India's 1974 explosion had, for example, benefited from those of a Canadian-built CANDU reactor. Even reprocessing technology no longer had to be indigenous; Brazil's purchase of West German reprocessing technology and equipment and the Shah of Iran's orders for the same from France were but the two most prominent confirmations of the fact that such was now available on the open market.

Delivery systems were becoming equally common. Throughout the 1960s and into the 1970s, most arms exports were of less than front-line quality. But as the 1970s progressed, increasingly sophisticated items came on the market.[10] The Shah, for example, had proved able to acquire sophisticated American fighter aircraft even before they became standard in the US Air Force; he received TOW antitank missiles before they became operational in the US Army; and he was given naval combat means not yet deployed by the US Navy. Sophisticated fighter-bombers were not just becoming available; it was a buyer's market, as sellers competed by offering ever longer payment schedules, ever more subsidized rates of interest.

The 1970s world of stagflation and recession in the developed countries saw moral sensibilities drowned by considerations of balance-of-payments impact. When Iraq became dissatisfied with Soviet delivery terms and spare parts availability, France stepped in to sell Mirage jets. When the Taiwanese feared that Washington's rapprochement with Beijing might entail restriction on sales to Taipei, they found a willing surrogate supplier in Israel. In general, though, there was no hesitancy, and no shortage of suitors—merely a matter of who offered the best dowry. Washington was quite happy to supply unrepresentative South American dictatorships, and so was Moscow (though the Soviet Union tended to be less successful on this continent, due to the strength of established American bonds). Brazil's conservative generals were happy to sell to Colonel Muammar el-Khaddafi's "revolutionary" government in Libya. Israel allowed itself to become the most important military supplier and quasi-ally of a South African regime whose elder statesman, former Prime Minister, B. J. Voerster, had declared during the Second World War that his party stood for the same principles as Germany's National Socialists, and indeed, that they differed in name only. Washington sold to self-proclaimed Marxists in Somalia; Moscow dangled luring deals in front of the royal house of Jordan.

Modern fighter-bombers were not the only possible nuclear delivery vehicles now becoming available. Certain missile types were also coming within wider reach. A select number of new nations were entering the ranks of those who had mastered ballistic missile technologies, as exhibited by the ability to launch satellites (Japan was one; India was another). But cruise missiles promised to have the larger impact. Cruise missiles, essentially pilotless airplanes or drones with air-breathing, continuously operating motors, were the modern descendants of Germany's V-1 rockets of the Second World War.

Because the V-1s had limited range, were slow, and could be shot down, postwar powers initially chose to pursue rather the technology of the V-2, which was a true ballistic missile. Ballistic missiles, akin to bullets fired from a gun or rifle, rely on the explosive thrust of propellant combustion stages, between which they float free-flight, free-fall, like a ball once thrown from the hand. The tremendous speed of these missiles (the boost velocity of an intercontinental-range ballistic missile is 5 km per second, or more) ensured that they could not be intercepted with traditional means, and that potential new-technology interception concepts would have to overcome awesome scientific odds.

But simpler cruise missile technologies also evolved, if only as a by-product of work focused on their more dramatic or illustrious cousins. Moscow's ship-to-ship cruise missiles in the late 1960s showed the way. By the late 1970s US sources testified that cruise missile range limitations could be overcome, and that the missiles could now be designed to meet intercontinental-range requirements. By this time satellite mapping techniques matched to on-board computer capabilities also allowed for terrain-hugging "contour guidance" and theoretically pinpoint "terminal targeting." Finally, while their propulsion mode continued to allow for easy interception if located, their low-altitude flight potential posed considerable problems for ground-based radars; their altitude would be too low for distant identification, thus ensuring that identification, if achieved, would be last-minute and transient, putting crucial premium on very fast reaction capabilities. The larger powers met the challenge by developing air-based look-down radars capable of distinguishing incoming drones from ground clutter, combining this with shoot-down air-to-air missile capacity.

The latter development was beyond reach of lesser powers, but then so were the technologies that allowed for ground-hugging cruise missile performance. A number of points need to be made. On the one hand, even the super-sophisticated airborne warning systems were coming on the market. 1981 saw the Reagan administration's determination to sell the US version, the so-called AWACS planes, to Saudi Arabia, and Britain's expressed willingness to step in with its only slightly less capable Nimrod planes in the event of congressional veto. (France was also in the market, with a putative system of its own.) More important, perhaps, was the fact that shorter-range cruise missiles, antiship, anti-air and antiground, were being proffered on the international arms sales market by the United States, the Soviet Union, France, Israel, Norway, and others. The basic technology was simple, and could be mastered by most developing nations, if they did not wish to buy. Intercontinental-range sophistication might remain a preserve of the superpowers for some time. But even here the technological obstacles were not so great as those associated with ICBM aspirations. And, in any case, most potential power aspirants identified the main perceived threats to their security as emanating from neighbors and less distant sources.

The most disturbing point related to the relative cheapness of cruise missiles, at least if one dispensed with the sophisticated accessories that would be demanded to penetrate the superpowers' multilayered defense systems. Cruise missile costs allowed one to contemplate acquisition of such numbers as might suffice to saturate opponent defense capacities. This prospect had less relevance in the superpower context. Either one of the mastodons could deploy a million cruise missiles, yet because their ballistic means already encompassed massive "overkill" potential, such deployment would serve only to provide even more ludicrously redundant capacity to restir the rubble; nonvulnerable and less vulnerable sea and air capacities would retain their absolute or relative immunity. But for weaker powers the low cost and associated saturation prospects of cruise missile technology opened new vistas.

The technical deterrents to "going nuclear," in particular those related to plutonium access and bomb delivery prospects, were by 1980 fast receding into the realm of the insignificant. At the same time new incentives for the embrace of nuclear potentials were becoming apparent.

One major incentive derived from the indisputable fact that the larger powers had not lived up to their side of the Non-Proliferation Treaty bargain. Signatories to that treaty had been solicited on the basis of a commitment by "the nuclear club" to pursue the goals of arms control and disarmament. The *quid pro quo*, explicitly spelled out by the treaty, was also its *sine qua non*. Yet even the more limited goal of arms control looked ephemeral in the context of the early 1980s; the upward spiraling of superpower armaments appeared impervious to arms control advocacies. The disillusioning process was furthermore complemented by increased evidence that the smaller nuclear powers perceived real political utility in the fact of nuclear possession. There was accumulating evidence that Britain's establishment, for example, had long seen the political impact to be derived from nuclear status as more weighty than narrow considerations of defense and military utility. In the words of *The Economist*: "The price of national prestige is higher than the cost of the Trident" (the $16 billion strategic submarine modernization program authorized by the Thatcher government).[11] There were many, in Tokyo and elsewhere, who ascribed China's permanent seat on the UN Security Council, and the continued denial of similar standing to Japan, to the fact that China's bomb capabilities weighed heavier in the scales than economic and other considerations.

The disappearance of previous technological deterrents and the emergence of political incentives encouraging pursuit of nuclear potentials were joined by still another set of circumstances that worked to the same end. The facts of ever-increasing resource competition, ever-worsening economic disparities between North and South, increased cynicism—and indeed desperation—that the established economic order(s) did *not* offer relief from systemic impoverishment, and that developed nations were *not* willing to countenance the required redistribution of assets all served to embed extremes of alienation. Volatile, explosive sores of potential conflict were multiplying like locusts.

This situation added yet another extremely dangerous element to constellations of traditional jealousy and threat. Analysts had long considered that Third World nations would be more likely to "go nuclear" as a result of localized considerations than as a consequence of superpower action and impact. In the words of one prominent author in 1975:

> Among nations outside formal alliances, the most dangerous situations related to those geographically and historically paired constellations of political adversaries: Israel and the Arab states . . . ; India and Pakistan; South Africa and black Africa, and—although prestige is probably a stronger motivation than the desire to threaten—Argentina and Brazil. No political prophecy will be made here, but it must be strongly underscored that an important dynamic factor is the chain-reaction pattern. If one link in the international concordance breaks the de facto non-proliferation bond, a grave risk of a run towards nuclear weapons is in the offing.[12]

Subsequent events increased the danger that subgroup dynamics would occasion spurts of proliferation. India's 1974 "peaceful nuclear explosion" spurred the Shah of Iran to authorize a dramatic nuclear reactor procurement program (a program thought by many to be enmeshed in weaponry aspirations), and encouraged resolute Pakistani emulation efforts. The former was aborted by the Shah's fall from power, but the latter was not detoured. By 1981 indications that Pakistan was approaching the testing stage increased India's determination to proceed to the stage of weapon systems deployment, which in turn, inevitably, strengthened Pakistani resolve. Similarly, there could be little doubt the CIA reports of Israeli nuclear weapons stocks provided a decisive spur to Libya's decision to help finance the Pakistani program (presumably with some expectation of investment return), and to Iraq's pursuit of nuclear reactor capabilities; nor could there be doubt that the Israeli bombing raid against Iraq's main reactor triggered renewed Arab determination to neutralize Tel Aviv's nuclear edge.

Elsewhere, the historical rival South American leadership candidates, Brazil and Argentina, had embraced ambitious civilian nuclear development programs, one resting primarily on German assistance, the other on Canadian, German, and Swiss aid.[13] The third member of the South American subgroup, Chile, was a nation whose rivalry with Buenos Aires was nearly as deep-rooted as was Brazil's, and considerably more volatile. Chile lagged in the race to nuclear power generation. But its technological potential was scarcely less than that of the others. It was clear that if either one of these three nations decided to pursue the nuclear weapons path that was within its grasp, the dynamics thereby unleashed would inevitably ignite parallel endeavors in the other capitals. The African subgroup of South Africa and black Africa showed less surface equilibrium, but even more focused and probably more implacable hostilities. The late 1970s/early 1980s evidence that Pretoria had developed, or stood on the verge of developing, nuclear weapons capability was without doubt a thesis that would generate its own antithesis, to borrow Marxist language.[14] Finally, in the Far East, there was the subgroup of Japan–Taiwan–South Korea. Some feared that Japan's nuclear abstinence was not immutable. For complex reasons China's nuclear explosion had not brought forth a Japanese echo, but Tokyo was not likely to greet a Seoul or Taipei explosion with equal acquiescence or equanimity. By the early 1980s both Taiwan and South Korea were posed so as to be able to bring a nuclear weapons program to fruition within a limited time period. Japan's relationship with both appeared more conflict-prone than did ties with Beijing. A range of potentially conflicting interests and flashpoints, including territorial differences, lay dormant and unresolved.

By the early 1980s the destabilizing potential of subgroup dynamics was further exacerbated by evidence of cross-group "fertilization." Concerning Israel, Taiwan, and South Africa, for example, the *New York Times* reported that "some intelligence and State Department officials who monitor the flow of nuclear technology and information are convinced that the three countries constitute the major players in an emerging club of politically isolated nations whose purpose is to help each other acquire atomic bombs."[15] The report continued: "Israel is said by intelligence officials to be assisting Taiwan in developing a rocket that could be used to deliver atomic warheads. . . . Intelligence officials report that Israeli scientists are working in South Africa on nuclear energy programs. . . . Scientists from Taiwan are said to be working in South Africa. . . . South Africa has become a supplier of uranium to both Israel and Taiwan in exchange, intelligence officials said, for critically important technology and training."[16] Similarly, there was evidence of "growing nuclear ties among other would-be atomic powers . . . . Intelligence aids say, for example, that Brazil recently agreed to provide Iraq with sensitive nuclear power technology obtained from West Germany . . . ."[17]

The danger of "a run towards nuclear weapons" that lay at the heart of the apparent unraveling of "the de facto non-proliferation bond" entailed particular perils over and above those immediately evident. The initial "strategic" capabilities of "Nth" states promised to be as primitive as early American and Soviet means had once been. Their missiles, if indigenously produced, were likely to be as faulty, as encumbered by so-called degradation factors, as Moscow's had been during the late 1950s. Their fuel technologies would also be in their infancies, as would command and control facilities. They would probably not in the first instance be able to construct hardened silos, nor would they be able to make full use of mobile land or sea (SLBM) deployment modes. Like early Soviet strategic potential, those of aspirant powers would be highly vulnerable and slow in reacting. Their nature would be such that they would have to be surreptitiously and premeditatively readied and fired in an initiating strike, or they would be in grave danger of being destroyed on the ground. Proliferation prospects thus promised to resurrect the "temptation syndrome" that had made the period of the late 1950s and early 1960s the most dangerous time span since World War II, and it promised to multiply the problem. Like the superpowers during the above-mentioned years, so newer powers would have to labor under less than adequate intelligence conditions (their satellite information-gathering potential would be putative at best), a situation that invited misperception and miscalculation. Finally, with more and more unstable subgroups approaching the nuclear threshold, sanity and prudence could no longer be guaranteed. The Idi Amin of the 1970s had not enjoyed access to nuclear arms; a successor of the late 1980s and 1990s might well.

Apart from the fact that inadequate delivery systems might in the end serve a lightning-rod purpose, there was the point that formal delivery systems might prove increasingly irrelevant. At the time of the Cultural Revolution, when Chinese hostility toward Washington was as unremitting and vitriolic as was their attitude to Moscow, and when their demonstrated bomb capacity was not welded to delivery systems capable of penetrating adversarial defenses, there was speculation that the Chinese might circumvent the impasse by smuggling bombs aboard merchant vessels or through commercial and nonorthodox channels. At times of minimal immediate threat, potential nuclear powers might have been deterred by inability to build or secure adequate delivery vehicles. The world of the 1980s was more tumultuous. By now delivery means could be acquired with greater ease. Yet such acquisition would still be costly, probably time-consuming, and certainly visible. One could not discount the possibility that nations that had secured limited bomb capacity through domestic, secret effort, possibly through theft, or even through loan, might employ such capacity in innovative and nontraditional modes, especially in response to a neighboring adversary's decision to pursue nuclear prospects.

Another specter from the past might also return to haunt power calculations. At the time of the most implacable Sino–Soviet hostility, during the late 1960s, there were those who suggested that China might one day be able to launch a single or multiple strike along trajectories that would invite the Soviet conclusion that America was the aggressor, thus sparking a US–Soviet conflagration. Future analogies, and their implementation, were not inconceivable.

The prospect of terrorist, and nontraditional state usage of nuclear means was made believable by the fact that nuclear bombs could now be made small and inconspicuous. The first hydrogen bomb had been 25 feet long. By 1980 the material for a plutonium bomb needed "to be no larger than a man's fist" to produce an explosion equal to 20,000 tons of TNT.[18] It could be transported in a briefcase, a handbag, even a pocket.

Terrorism is perhaps less useful as an object of prime concern, if only because the steps that can be taken to guard against it are limited (essentially they are restricted to better security at bomb depots, better security at fuel storage and transport facilities, and more stringent travel controls).[19] Also, while terrorism might have horrendous implications for a particular locale, the worst-case devastation is likely to be less widespread than conjured up by state designs, and the dreaded escalation snowball is not likely to begin rolling.

The possibility of terrorist assault against a nation or community of nations through nuclear blackmail is not the only aspect of the fear of terrorism. Some thoughtful observers of world events have seen greater reason to fear "the danger that the weapons states will themselves become terrorist, and turn their terror against their own peoples": the Third World was replete with cases of increasingly embattled, unrepresentative, and sometimes desperately repressive governments; the developed world, meanwhile, witnessed assertive government actions to choke off and indeed preclude debates on matters of nuclear policy.[20] The fear of nonstate terrorism might actually serve the interests of state terrorism, by justifying policies that restricted or mocked civil rights norms and ideals; it invited the misuse of power. But while civil libertarians might despair, and the quality of life might suffer, these also were not trends that in and of themselves affected proliferation and escalation prospects—at least not directly.

As concerns proliferation, with its attendant risks of conflagration and escalation, the control outlook was less than reassuring. Previous deterrents had been whittled away. Incentives to "go nuclear" were on the rise. And there was no authority on the horizon that appeared capable of intervening, to reverse these ominous trends. UN authority remained a mirage. The writ of its moral suasion was limited; it tended to founder on the shoals of chauvinism and sectarian interests. And its powers of enforcement were at the mercy of governments jealous of UN authority and disinclined to concede on matters of sovereignty.

The physical means required to halt proliferation trends remained the preserve of the superpowers. In a more rational world the ultimate imperative of self-preservation might have been seen to dictate collective action. And such collective action might well have been supported, however grudgingly, by most of the world community. Many would probably have sympathized with Chinese fears of "superpower condominium," and what it might entail, yet the alternative appeared fraught with even worse dangers.

In the world of the early 1980s, however, the argument was academic. The bitter confrontational politics of the superpowers precluded joint action. Control aspirations had to rely on more restricted damage-limiting concepts. In areas clearly within their spheres of influence and exempt from the immediate pressures of superpower rivalries, the dominant powers obviously had a potential for leverage. Within this restricted context Moscow's record was good. Soviet safeguard standards, in terms of allowing nuclear technological, informational, or other access to foreign hands were exemplary (or, at least, had been exemplary since 1961).[21] The Western record was more mixed. There was some reason to believe that American disapprobation, expressed through diplomatic and perhaps economic channels, had stayed Seoul's hand, and that pressure on Islamabad (in the late 1970s) might have slowed Pakistani efforts. But all three nations that had joined, or were thought to have joined, the nuclear club during the 1970s—India, Israel, and South Africa—owed their success to the availability and suitability of waste products from Western-provided reactors. And the Reagan administration's 1981 reversal of its predecessor's decision to impose aid sanctions on Pakistan (as a riposte to Islamabad's refusal to disavow nuclear ambitions) appeared to signal the effective demise of Washington's antiproliferation stance.

The brakes on proliferation seemed ever more worn, just as the engine was revving up. In conclusion, one might just note that if Third World proliferation prognoses proved right, then even nations whose abstinence was one of principle might feel forced to reconsider. The actuality of 20, 30, or more nuclear-capable but economically alienated developing powers would undermine even the moral abhorrence of Scandinavian and like-minded nations. Pillars of UN peace-keeping ventures, like Norway, Ireland, and Canada, were furthermore likely one day to have to face the prospect that state or nonstate actors involved in these scenarios would have access to nuclear weaponry. Once proliferation passed a certain stage it might no longer be a question of who could "go nuclear," but rather of who could afford *not* to.

NOTES

1. *The Military Balance 1981–82,* IISS, London, Fall 1981.
2. *The Annals* (of the American Academy of Political and Social Science), March 1977; and see, e.g., William Epstein, "Ban on the Production of Fissionable Material for Weapons," *Scientific American*, July 1980.
3. From "The Genie Escapes," *Toronto Globe and Mail*, 25 April 1981, reprinted in "The Bomb—Beyond Control" report, *World Press Review*, August 1981, p. 38.
4. The US government's attempt to stop the *Progressive* journal from publishing an article containing nuclear bomb information culled from public sources was abandoned after similar articles were printed by other magazines and newspapers.
5. Stockholm International Peace Research Institute, *The Nuclear Age*, Almqvist & Wiksell, Stockholm, 1975, pp. 79–82.
6. Epstein, "Ban on the Production."
7. See, e.g., Mason Willrich and Theodore Taylor, *Nuclear Theft: Risks and Safeguards*, Ballinger, Cambridge, Mass., 1974.
8. "Nuclear Proliferation—Prospects in the Middle East," interview-article, *The Search, Journal of Arab and Islamic Studies*, Summer 1981.
9. To cite but one typical example: between 1 October 1979 and 31 March 1981, 25 pounds of plutonium, enough to make two or three nuclear bombs, disappeared from the Savannah River Plant in South Carolina. A US Department of Energy spokesperson suggested that the disappearance could be "accounted for by stuff that's stuck in the piping and plumbing and measurement imperfections," but admitted that the amount "seems to be more than Savannah River had missed this time last year;" the official noted that "there are differences many times in shipper and receiver measurements." UPI report, *Miami Herald*, 1 August 1981.
10. Anthony Sampson, *The Arms Bazaar,* Hodder and Stoughton, London, 1977.
11. *The Economist*, 28 February 1981 (review of David Dilks, ed., *Retreat from Power: Studies in Britain's Foreign Policy of the Twentieth Century*, 2 vols., Macmillan, London). The price estimate, £8 billion (about $16 billion at then-current exchange rates) was made by David Greenwood, Director of Aberdeen University's Centre for

Defence Studies; see the (London and Manchester) *Guardian Weekly*, 20 September 1981.

12. Alva Myrdal, *The Game of Disarmament—How the United States and Russia Run the Arms Race*, Pantheon Books, New York, 1976; see Spokesman edition, London, 1980, p. 181.
13. Note, e.g., "Brazil Pushes Nuclear Effort in Spite of U.S.," *New York Times*, 21 August 1980.
14. Nigeria and Zaire both appeared on the 1980 *Scientific American* chart of nations capable of "going nuclear" within a decade (other African nations on the list, aside from the Republic of South Africa, were Libya and Egypt); see *Scientific American,* July 1980. As concerns South Africa's nuclear capabilities, see *South Africa: Time Running Out,* The Report of the Study Commission on US Policy Towards Southern Africa, University of California Press, Berkeley, 1981, pp. 251–253.
15. *New York Times*, 28 June 1981. The particular instability occasioned by the nature of these three "club" members, by the fact that the governments of each pursued apartheid-type policies that excluded ethnic majority (or potentially majority) groups from equal access to levers of power, was added cause for unease.
16. Ibid.
17. Ibid.
18. *Comprehensive Study on Nuclear Weapons*, United Nations Report to the Secretary General, New York, September 1980.
19. See Louis Rene Beres, *Apocalypse: Nuclear Catastrophe in World Politics*, University of Chicago Press, Chicago, 1980; see also Nigel Calder, *Nuclear Nightmares: An Investigation into Possible Wars*, Viking Press, New York, 1980, p. 64.
20. E. P. Thompson, "The END of the Line," *The Bulletin of the Atomic Scientists*, January 1981: "Sometimes, as Zuckerman has noted, 'the rules of official secrecy are exploited, not because of the need for security, but to promote partisan policies' as between competing interests within the state bureaucracies. More generally, it is part of the overall exercise in manipulating domestic public opinion."
21. Khrushchev's alleged memoirs (the manuscript was clearly 'retouched' in places by Western editors) contained the assertion that Moscow had "kept no secrets from them [the Chinese]. Our nuclear experts cooperated with their engineers and designers who were busy building an atomic bomb. We trained their scientists in our own laboratories" (*Khrushchev Remembers—The Last Testament*, ed. and trans. Strobe Talcott, Little, Brown, Boston, 1970, p. 268). Chinese authorities, however, assert categorically that Moscow refused to aid their nuclear program, and that this refusal was in fact one of the important reasons for the 1961 estrangement between the two countries (as a member of a Canadian delegation to Beijing, in 1974, this author was given a long treatise on the matter by a member of China's Foreign Affairs Commission).

Chapter 6

The Arms Dynamic: Can the Genie Be Controlled?

1980: World military budgets spiraled past the $500 billion mark, rushing toward the $600 billion plateau. The United States alone spent more than *twice* as much on just basic training (boot camp) as the total sum available for education for South Asia's 300 million school-age children.[1] One-fifth of 1 percent of the world military budget would have paid for 100,000 elementary school teachers, at American salary scales.[2] Less than 5 percent would, as previously noted, have sufficed to overcome the cancerous horror data of world illiteracy, starvation (which kills one-third of the world's children before the age of 5), and unnecessary disease, malformation, and death (such as the more than 25,000 *daily* deaths caused by avoidable water-borne diseases).[3] Yet, with existing arsenals already averaging out to 30 tons of TNT for every man, woman, and child on earth, arms budgets continued to escalate, at an ever faster pace.

There were of course many varied and complex reasons for the apparently inexorable increase in world armaments. The single factor that required primary focus, however, was that of "the military-industrial complex." By 1980 the military-industrial complex had become a juggernaut with awesome political and economic influence in both superpower capitals—as well as in the capitals of a mushrooming number of lesser states.

The history of the military-industrial complex in the United States is illustrative (its Soviet analogue is discussed later). Prior to World War II the military-industrial sector of American society was of only peripheral importance. Military spending was no more than about 1 percent of the GNP. As in other Western democracies, the military establishment was kept at arm's length by the civil power. This was to change quite dramatically during the postwar decades.[4]

In his farewell address (January 17, 1961), President Dwight Eisenhower was moved to warn that "a permanent armaments industry of vast proportions" had become a fact of American life: ". . . three and a half million men and women are directly engaged in the defense establishment. We annually spend on military security more than the net income of all United States Corporations. . . . This conjunction of an immense military establishment and a large arms industry is new in the American experience. The total influence—economic, political, even spiritual—is felt in every city, every State house, every office of the Federal government. . . . The potential for the disastrous rise of misplaced power exists and will persist."[5] Yet the process had barely begun.

The "immense military establishment" of Eisenhower's time might already have been able to make itself "felt in every city, every State house, every office of the Federal government," but its influence was that of an increasingly powerful and assertive outside lobby. It was still not part of the club. The Nixon presidency proved a watershed in this regard. By the late 1960s and early 1970s the conscious divorce of military and civil power had given way to incestuous cohabitation. Managers and officials of military-oriented industrial giants began to be invited to serve in Pentagon positions, from which they would later return to their parent companies. The complementary pattern of Pentagon officials retiring to second careers on the boards of military-industry concerns also became the norm. Pentagon procurement decisions were increasingly made or influenced by people with direct financial stakes in the outcome of those decisions.

The pervasiveness of military industry's direct influence within the Pentagon and of its indirect (lobbying) authority in Congress is best summed up by one startling fact of the postwar era: neither the civilian Department of Defense nor the US Congress ever publicly recommended cancelling a military procurement program. President Carter's unique cancelling of the B-1 bomber program was of course made by the White House, in response to the near-unanimous skepticism of those segments of the scientific community not beholden to the air force. But the suggested assertion of independent strategic decision-making power was undercut when his successor, Ronald Reagan, reversed the decision. There was no question that some procurement policies were misguided:

> Consider the reported weapon reliability "at the target" of the following small sample of cases: A-6A "Intruder" fighter bomber tracking radar—40 percent reliable; F-111A terrain-following avionics—38 percent reliable; F-106 Delta Dart fire control system—4 percent; sidewinder missile—63 percent; Sparrow missile—46 percent; Antiradiation missile—50 percent; Phantom F-4 guidance systems—33 percent reliable! Only a third of the F15 "Eagle" air-superiority fighters, at $20 million each, are ready for combat at any one time. . . . The highly touted F-14 and its Phoenix air-to-air missiles are so unreliable and ineffective, even in peacetime tests, that it would be embarrassing to reveal their actual performance record.[6]

Other examples of weapon systems not performing as expected are legion. The cruise missile whose sophisticated guidance technology can be thwarted by smoke and outsmarted by snowbanks (see Chapter 3) is but one of the all-too-many items that could be added to the litany. Actually, all major strategic weapon systems have ended up both costing more than originally projected and incapable of fully meeting promised performance standards. Slippage has become integral to the system. Yet this skewed "norm" is not here at issue. The point is rather that the failure to abort funding even in cases of blatant program failure torpedoes the postulated relevance of foreign threat and cost-effectiveness prognoses—these *prima facie* considerations are clearly superseded by the inner-institutional socioeconomic dynamics of the military-industrial behemoth.

The potency and political impact of these dynamics is evidenced also in the extent of Washington's involvement in the military-industrial complex's foreign sales ambitions. Government involvement in the arms trade, both through active sales promotion and through financial subsidies, has increasingly become a universal phenomenon. In neutralist Sweden, as in the "born-again" administration of President Carter, philosophical opposition to arms proliferation had to defer to the domestic politico-economic clout of arms sales protagonists. More startling still was the evidence in the 1970s of cases in which US government support for the latter's aspirations went well beyond the normal techniques of promotion and export subsidy.

Canada's billion-dollar Lockheed purchases in the early 1970s are a case in point. Lockheed had been reeling from mismanagement, compounded by foreign scandal. Twice the corporation had been given congressionally guaranteed loans. But now, in the wake of foreign order cancellations caused by its admitted bribery of foreign officials (in itself a common enough practice), Lockheed again tottered on the verge of bankruptcy. Congress refused to sanction a third bail-out. Because the political consequences of the company's possible collapse was judged to be intolerable, the US administration had no alternative but to pressure an ally, through whatever means were necessary, to step into the breach. The planes were far from ideal for Canadian needs. Their potential for mid-Atlantic anti-submarine warfare met requirements of an epoch that was rapidly receding into history (Moscow was now building intercontinental-range SLBMs that could be fired from home waters, dispensing with the need to traverse seas dominated by NATO ASW capabilities); and the planes' general Arctic, eastern, and western seaboard surveillance capabilities were atrophied by the fact that their immense cost precluded the purchase of sufficient numbers—at the end of the day their maintenance, runway, and fuel requirements translated into the dismal fact that they would not substantively improve on the limited surveillance capabilities of Canada's past and aging means.[7] To the contrary, the exorbitant cost of the Lockheed purchase meant that Canada could not pursue the capability-enhancing programs that might have been within its reach (such as cheaper short take-off and landing planes, and fast missile-armed motor torpedo boats). Canada's purchase of Lockheed weapons was not the only case of arm-twisting by Washington to effect weapons sales for reasons other than strategic logic. It was but the most obvious example.

By 1975 world military expenditures already exceeded world expenditures on education and health; they already equaled the combined gross national products of all the countries in Africa, the Middle East, and South Asia.[8] By 1978 the United States and Western Europe were spending more than $20 billion on just military research.[9] Two years later, foreign military sales exceeded $30 billion per year (African weapons import alone had increased 21 times between 1969 and 1978).[10]

By the late 1970s half of the world's scientists had become employees of "the military-industrial complex." In 1975, 400,000 scientists and engineers were "employed in military pursuits in the advanced countries;" by 1980 US and Soviet numbers alone totaled more than 750,000, and there was no sign whatsoever that the upward momentum was abating.[11] The performance characteristics and cost of weaponry accelerated apace. The price of American weapons systems had multiplied 200 times since World War II, a testimony to previously undreamed-of sophistication—and cost-plus liberality.[12]

The number of people paid directly or indirectly by defense ministries had passed 100 million.[13] The real number of dependents was probably far greater still, if one counted collateral dependence. Branch plants of military industry, military bases, and the like also provided financial sustenance to the proprietors and families of nearby laundromats, snack bars, tobacconists, and other houses of good and ill repute.

World power establishments appeared to have accepted the proposition that the health of their military-industrial complexes had become crucial to the health of national economic infrastructures, and to political fortunes. The Lockheed case testified to the American situation. The European situation was not very different. In France 88 percent of all government-funded "civilian" R&D (research and development) went to military-oriented aircraft, space, nuclear energy, and computer industries. In West Germany, the figure was 87 percent; in the Netherlands, 67 percent; in Britain, 86 percent.[14]

Independent data strongly indicated that the proposition was false. Indeed, in-depth historical analysis provided powerful testimony to the effect that there was in fact an inverse relationship between military expenditures as a percent of GNP and a nation's annual rate of growth in productive capacity.[15] The possibilities conjured up by the million or so world "military" scientists redirecting their energies to productive industry are near limitless.[16] The ultimate visions of disarmament advocates were of course utopian. Reasonable observers were bound to concede a plethora of legitimate security concerns. But the demonstrable fact of continued funding even of programs whose products were unmitigated failures in both military and economic terms bespoke too automatic a deference to the burgeoning power of the military-industrial bureaucracy.

Too many seemed to have forgotten Eisenhower's admonishment:

> The craving for absolute security. . . can bankrupt itself, morally and economically, in attempting to reach that illusory goal through arms alone. The Military Establishment, not productive of itself, necessarily must feed on the energy, productivity, and brainpower of the country, and if it takes too much, our total strength declines.[17]

By the early 1980s:

> Such huge resources—manpower and money—have been devoted to military science for so long that the momentum of military technology is now well nigh irresistible. The military-technological tail wags the political dog.[18]

The Soviet analogue to America's military-industrial complex was a beast of different antecedents, yet remarkably similar present characteristics. Moscow's military-industrial conglomerate was born and nourished in a rather different philosophical and cultural environment. The Western tradition had viewed the military as something to be cultivated but never embraced. The Russian tradition, as Eastern traditions in general, was always more integrative. The cultural bias was reinforced by the Clausewitzian element in Leninism and Bolshevism (the view that state power was a totality of economic, bureaucratic, military, ideological, and other means, with the particulars of their interplay, distinctive emphases and de-emphases determined by pragmatic, evolving, and changeable assessments of relative utility and purpose). The predilection was further strengthened by the realities of early Bolshevik rule. The Leninists had of course not swept the country in 1917. The Bolshevik success in Petrograd (Leningrad) was followed by success in Moscow, then other towns and other areas. But if one charted their progress by red shadings on a map, one would see only a slow spread of often isolated red dots, the gradual establishment of secure interconnecting arteries, and a by no means inexorable outward seepage of shaded red patches; it was 1924 before the country was uniformly red, before the last civil war antagonists had been defeated and the last hostile foreign troops withdrawn.

It was a battle against exorbitant odds. Hence there could be no question of separating political and military leadership. The political leadership was the military leadership; political supporters were *ipso facto* the front-line troops. From the birth of the new "Socialist Republic" the military was not just an integral part of the Party; in some ways it was the Party. World War II saw Party leaders in fighting roles; the Party membership at large sustained the greatest losses of any single societal group.

Extrapolating from their own environment and their own socialized preconceptions, Western analysts have often postulated severe military-political friction in the USSR. But even the military purges of the late 1930s did not represent a Party assault on the military hierarchy; rather, it was part and parcel of Stalin's assault on independent pillars of the Party-military establishment. (Here also, it was the Party that suffered the most, having been decimated to a greater degree than any other group or conglomerate). Western group analyses have tended to be too facile. The leadership of the Party and of the armed forces remain parts of the same composite. The leadership of both served together not only in wartime, but on district military councils before and after World War II; they continue to serve together in the Party Central Committee, the Stavka or Higher Military Council attached to the Council of Ministers, and in the Politburo. "Military" men may engage in factional politics of various kinds and stripes, but they do so as representatives of (often conflicting) Party tendencies, not as outside lobbyists.[19] There is obvious congruence between basic military and Party aspirations, at home and abroad.

Professional empathy wrought through decades of intimate association reinforced the dictates socialized through decades of inferiority and insecurity. There is no evidence that "Party" leaders ever vetoed a weapon system deemed necessary by the armed forces hierarchy. The demobilizations of the early postwar period and the late 1950s and the mid-late 1960s hiatus in BMD deployment were consensual decisions. (The first responded to the supreme reconstruction requirement occasioned by the wartime loss of 70 percent of the industrial base; the second was occasioned in part by the post-Stalinist determination to divert increased funds to consumer goods, a determination encouraged by the overblown expectations that surrounded early ICBM achievements; the third derived from the admitted inadequacies of then-available defense technologies.) There was no evidence of discord on the basic premise that Soviet defense capabilities must be improved, and that inferiority vis-à-vis capitalist power groupings was an intolerably dangerous condition.

The license this accorded Soviet defense planners meant, essentially, that their designs could be pursued in an environment devoid of basic criticism. As in the increasingly analogous American case, this in turn inevitably led to at least some instances of program fundings that clearly would not have survived independent strategic and cost-effectiveness analyses, but which owed their existence solely to bureaucratic needs integral to the military-industrial complex. The astoundingly long life accorded to a tank gun loading mechanism that could be operated safely only by left-handed personnel, and to a jet liable to shake itself to pieces if flown anywhere near its supposed top speed, were but two examples of the kinks inevitably spawned by organizational dynamics unfettered by outside critique.[20]

The bureaucratic phenomenon that was the modern "military-industrial complex" had developed into a beast with independent momentum and dynamism. It had acquired a particular potency that emanated from the very nature of the organization, and that was largely immune to the vicissitudes of outside threat and cost-effectiveness projections. It is crucial to note that bureaucratic considerations and politics had come to dominate developmental priorities and decision, to the detriment of, and sometimes even to the exclusion of, value-free calculations of pro forma rationales.

Nevertheless, attention must also be paid to other factors fueling the engine of arms inflation, if only because they have served to cocoon the military-industrial complex from outside criticism. One of these factors, and one that promised to fuel and bedevil the armament programs of "Nth" states just as it had those of the superpowers, was the concept of "lead-time." This is the time required to research and develop a new weapon system. The lead-time for a modern weapon system averages out at 10 or more years. This means that one must today plan (and lock oneself into) programs designed to meet threats that may eventuate 10 to 20 years down the line. Since one does not always know what choices an opponent may make among the many options for weapons improvement and development that present themselves, and since one cannot know which technological "gleams in the eye" the opponent will be able to master, and which will prove elusive or illusory, there is always an inherent and very real "need" to allocate for efforts far out of proportion to what history may justify.

The lead-time concept is of course inextricably intertwined with that of "worst-case planning." It is in the nature of defense planning that one maximizes the opponent's ability to overcome difficulties while minimizing one's own. The tendency was typified by a US government presentation to Congress on the subject of relative Soviet–American naval strength: a rather hefty portion of theoretically available American vessels were excluded from the ledger, with no concomitant dilution of Soviet numbers; a challenge on this score was answered by the statement that US figures represented the ready-for-combat category, while Soviet figures had been left all-inclusive because exact percentages of Soviet vessels being repaired or otherwise laid up at any one moment were not available.

Two other examples, from the early 1960s, spring to mind. Secretary of Defense Robert McNamara and his "whiz kids" were disdainful of existing ballistic missile defense technologies and refused to budget for an American effort in the field. They were scathing in their comments on the primitive defense system that Moscow nevertheless persevered with. Yet, although extremely skeptical about the effectiveness of the Soviet endeavor, they nevertheless felt compelled to order countermeasures based on the assumption that the Soviet system might work after all. Immensely expensive programs of "penetration aids" and multiple-warhead and "MIRVing" developments were authorized to counter what was claimed to be ineffectual. The reverse side to this coin was found in sometimes ludicrous extrapolations of what could go wrong with offensive missiles. Early such missiles were, as will be recalled, primitive and faulty, afflicted by a number of so-called degradation factors (categories of faults, some due to design or production mistakes, some to transport, storage, firing, and in-flight inadequacies and phenomena). By maximizing everything that could conceivably go wrong, one arrived at very limiting estimates of effectiveness. Since policy planners at the time were unaware of the debilitating effects of "fratricide" (the first warhead explosion, whether in the air or off or on target, will itself incapacitate accompanying means), the obvious answer appeared to lie in a quantum leap in numbers, the multiplying of missiles assigned to each target.

Lead-time and worst-case considerations cannot be faulted for the continued nurturing of procurement programs that have proved faulty and inadequate. But they are often presented as the most important causal factors behind the arms race. Arms race theories are usually seen in action-reaction terms, both at the macro level of opposing force conglomerates and at the micro level of particular-unit and lower-unit weapons developments. One oft-quoted illustration postulated that early Soviet ballistic missile defense efforts were spurred by the awesome numbers of the American offensive arsenal; Soviet BMD testing was then said to have led to American development of decoys and penetration aids; this in turn supposedly spurred Soviet radars capable of distinguishing the real from the false (differently composed bodies fall through the atmosphere at different velocities), which was thought to have resulted in ever more sophisticated American decoy designs, culminating in the practice of multiple free-fall cluster warheads (MRVs). The Russian response, according to this scenario, was to circumvent MRV saturation prospects by designing long-range intercept missiles that could reach the incoming booster rocket before it released its warheads; finally, the Americans developed independently targetable and maneuverable warheads that could be separated from their booster beyond antiballistic missile reach; and we arrive at presumed Soviet research into land- and space-based laser and particle beam potentials, the specter of which has secured extensive funding for American ambitions in the field.

Arms race action-reaction theories tend, as previously noted, to be too simplistic. Chapter 3 presented testimony by specialists privy to American decision-making councils, to the effect that Soviet prowess, actual or potential, was often incidental to force deliberations. The early 1960s decision to set a quota of 1054 missile boosters was, for example, clearly not affected—at least not primarily or significantly—by projections of Soviet strength. It is not that the action-reaction syndrome never operated in real life. Action-reaction theories were prominent in many facets of strategic debate, and the reality and implications of lead-time and worst-case calculations were beyond dispute. Some observers were less than impressed with the military-industrial complex's basic level of scientific-technical competence.[21]

Others thought that analyses of the opponent's scientific potential left much to be desired.[22] And one might indeed argue plausibly that immunity from the rigors of outside scientific and academic inquiry inevitably had a debilitating effect on threat projection and response design alike. Notwithstanding the seriousness of this critique, however, it was incidental to the main problem—which derived from the fact that affirmed legitimizing premises did not suffice to explain the scope, nature, and *modus operandi* of the military-industrial complex.

There are those who take refuge in the concept of "technological imperative." The Protestant ethic in the West and the Bolshevik creed both place progress on a pedestal. Both are said to harbor a cultural predisposition favoring progress for its own sake, an inherent inclination to favor pursuit of such vistas as are suggested by evolving technological potential. The argument can be made quite compelling. But while no doubt a factor, this concept also falls short of providing sufficient explanation for the character of today's military-industrial complex.

The more this conglomerate and its workings are studied, the more it becomes apparent that a major part of the answer can be provided only through resort to organizational theory and the fundamental sociopolitical concept of "bureaucratic politics." A grotesquely distorted caricature of its (very) legitimate self, the military-industrial complex may have become a cancer within the modern body politic. But, if the analogy is apt, the point must be made that it has obviously spread too far to be easily excised; hopes for arrest and control must rest with long-term, patient, and concentrated treatment. In other words, it is not merely a matter of questioning the scientific soundness of particular procurement decisions or of producing more reasoned estimates of Soviet potentials, and it is not merely a matter of individual politicians and their will and purpose. A more critical circumscribing of military-industrial power and purpose would demand major change in reigning sociopolitical attitudes, and it would demand more purposeful and sustained involvement on the part of academic scientists—academic scientists with access and influence beyond their present dreams, or even desires.[23]

Yet the international environment of the 1980s clearly made the task of reining in established military-industrial complexes more difficult than ever. All the perceived legitimate rationales that had fostered the growth of these complexes remained vibrant. Furthermore, the 1970s and early 1980s witnessed the emergence of a range of new conflict-prone phenomena and scenarios. It was no longer a matter just of traditional threats and rivalries, compounded by the technological-conceptual dictates of lead-time and worst-case requirements, and the luring but otherwise neutral temptations of new scientific breakthroughs and vistas.

A new Pandora's box of escalating conflict potential had arrived on the world scene—a Pandora's box containing ever-increasing North–South disparities, resource and energy scarcities, stagflation, and the failure of traditional economic theory and management, of the old "economic order." The destabilizing potential inherent in any and all of these, and in their sometime interrelation and interdependence, was awesome. It was all the more awesome when seen in conjunction with the fact of ever wider and easier access to nuclear means and technologies.

The oil embargoes and dislocations of the 1970s were obviously but the harbingers of an era of ever more acute resource scarcities. The continuing resource profligacy of Western developed economies, and especially of the United States (notwithstanding the limited conservation efforts of the late 1970s and early 1980s, the US economy remained far more wasteful, of energy in particular, than most of its smaller rivals), jarred against the needs and ambitions of developing economic infrastructures, and the fact and danger of depleting resources. New geological discoveries might postpone crises, but potentials remained finite.

The squeeze was exacerbated by the incontestable fact that the pioneering Organization of Petroleum Exporting Countries (OPEC) typified a more general determination on the part of exporters to maximize the politico-economic benefice of their remaining bounty, through the formation of similar cartels. Developing countries had long suffered from resource exploitation. A number of minerals had, like oil, suffered from outside (essentially Western) economic price and market control, with the net result that their value had suffered in comparison with that of (Western) manufactured products; the postwar price of the former had not been allowed to keep up with the inflation-compensating and profit-ensuring price increases of the latter. Western advantage had previously been perpetuated through financial divide-and-conquer tactics and through all-too-credible threats of intervention. OPEC's very existence demonstrated that the first method could be countered, and that the second could no longer be pursued with the license of yesteryear. OPEC demonstrated the feasibility of collective action. It also increased the need for action, since it coincidentally served also to wreak additional havoc with other developing nations' trade balances. The relative failure of cartel aspirations during the late 1970s, and the problems encountered by OPEC itself, testified to the enduring politico-economic clout of the developed world, and to the difficulties that still beset ambitions for equitable resource control schemes. Nevertheless, long-term trends promised to strengthen the force and urgency of cartel advocacies.

These trends spotlighted another problem area. The minerals now becoming scarce encompassed items of high strategic value. This meant that the implicit threat to developed civilian economic infrastructures was paralleled by increased threat to developed nations' military might, a threat striking at the heart of their military-industrial leviathans. The threat posed a real and long-term, increasing danger to the military-industrial complex's ability to meet its primary obligations, the obligations arising from traditional rivalries (first and foremost, the US–Soviet confrontation). Since these obligations constituted the core of military-industrial legitimacy, the constellation served *ipso facto* to legitimize and indeed give priority to assertive actions to counter the new threat.

America's politico-military posture was most affected. The USSR was less autarchic than in earlier years but remained far more self-sufficient in strategic minerals than was the US. 1981 statistics showed that America depended on outside suppliers for between 50 and 100 percent of annual requirements for 20 essential categories of strategic minerals. 100 percent of US columbium needs were met by Brazil, Thailand, and Canada; 100 percent of mica (sheet) supplies came from India, Brazil, and the Malagasy Republic; 100 percent of strontium was delivered by Mexico and Spain; 98 percent of the manganese came from Gabon, Brazil, and South Africa; 97 percent of tantalum from Thailand, Canada, Malaysia, and Brazil; 97 percent of cobalt from Zaire, Belgium, Luxembourg, Zambia, and Finland; 93 percent of bauxite and alumina from Jamaica, Australia, and Surinam; 92 percent of chromium from South Africa, the USSR, Zimbabwe, and Turkey; 91 percent of platinum group metals from South Africa, the USSR, and the United Kingdom; 84 percent of asbestos from Canada and South Africa; 82 percent of fluorine from Mexico, Spain, and South Africa; 81 percent of tin from Malaysia, Bolivia, Thailand, and Indonesia; 77 percent of nickel from Canada, Norway, New Caledonia, and the Dominican Republic; 66 percent of cadmium from Canada, Australia, Belgium, Luxembourg, and Mexico; 62 percent of zinc from Canada, Mexico, Australia, Belgium, and Luxembourg; 61 percent of potassium from Canada, Israel, and West Germany; 61 percent of selenium from Canada, Japan, Yugoslavia, and Mexico; 57 percent of mercury from Algeria, Canada, Spain, Mexico, and Yugoslavia; 54 percent of gold from Canada, Switzerland, and the USSR; and 50 percent of tungsten from Canada, Bolivia, Peru, and Thailand.[24] The vulnerabilities and consequent policy dynamics suggested by this list were made more combustible by the fact that existing stockpiles showed "shortages and imbalances in several key categories."[25]

While acknowledging the obvious specter that this conjured up as concerned Moscow's potential to affect delivery schedules and security through crude military means, prime attention should probably be focused elsewhere. The problem is perhaps best typified by central and southern Africa. Many of the critical suppliers on the above list are located in this area. The extreme American concern occasioned by the nature and extent of existing and projected supply dependence, and by the perception that "indigeneous instability, in some cases fomented and fanned by outsiders, poses serious political and military challenges,"[26] created an inevitable bias in favor of measures that would improve Washington's military posture. It also added urgency to the bias. It was therefore perhaps inevitable that Washington would focus on reviving and extending military-strategic ties of the past. That meant the Republic of South Africa, in southern Africa (and Israel in the Middle East). The perceived urgency of supply "threats" thus drove Washington to apparent embrace of a regime that, because of its official racism and because of its continued occupation of Namibia, had become a pariah in the rest of the continent (and one might note that continued Israeli occupation of Arab lands had had an analogous impact in the Middle East).

The focus on short-term rather than long-term security requirements had two immediate consequences. First, an alienated black Africa became more receptive to Soviet blandishment. South Africa's aggressive stance was seen to legitimize (and necessitate) the Soviet and Cuban presence in Angola, and elsewhere, and not just in black African eyes. One typical European commentary, titled "When Russia fights for freedom," ran as follows:

> ". . . the simplistic antithesis of the Free World versus Totalitarianism has never looked less apt than in Southern Africa, where it is the West which largely reinforces minority rule and the East which is allied with the values of majority rule and national dignity.[27]

The American stand thus facilitated Moscow's ambition to establish a Soviet presence in the area. It permitted Moscow to present its own *realpolitik* interests as congruent with the highest ideals of Western morality and Western political philosophy. By allowing Moscow to cloak its chauvinist and particular policy goals in the rhetoric of the Fathers of the American Revolution, it legitimized Soviet presence and sowed the seeds of more profound long-term black African alienation from American purpose.

But even if Moscow and/or its allies had chosen not to try to pick this ripening fruit, or even if they ultimately gagged on it, long-term American aims still were ill served. The alienation of black and "neutral" opinion that attended America's quasi-alliance with Pretoria had force of its own. It fanned anti-Western and anti-"Northern" economic resentments. And it increased prospects for black recourse to arms, against South Africa, and against the economic interests of its apparent ally/allies.

The anti-apartheid, antiracism, and anticolonial struggle which promises to continue to engulf southern Africa may be seen as but one aspect of the more general problem of North–South tensions, yet the fact that the struggle rages in an area that harbors some of the world's rarest and most vital strategic minerals ensures that it will inevitably affect also the immediate calculations of military planners in the United States, the USSR, Europe, and elsewhere. The problem of depleting resources—be they oil, minerals, or protein—may well be the single most likely generator of future armed conflicts. Other traditional rivalries—be they derived from irredentist aspiration, ideology, or whatnot—appear more manageable.

The Africa of 1980, the Africa that had absorbed a 21-fold multiplying of arms imports in the span of a mere decade, already included one presumed nuclear-capable state, South Africa. Other African states were among the proliferating number of (increasingly alienated) nations capable of procuring nuclear means in the not-too-distant future.[28] The indigenous maelstrom of irreconcilable black-apartheid friction was a prescription for disaster even without the involvement of outside powers. Dynamics internal to the subgroup of nation-states in this area were clearly far more likely to determine possible embraces of the nuclear path than were considerations spawned by superpower actions or reactions. In that sense, southern Africa was a prime example of the growing number of interstate subgroups, whose policy options centered on member interactions rather than on global phenomena.

In conclusion, it is necessary to observe that none of the forces driving current arms proliferation trends (nuclear and conventional) can be seen in isolation. Most are intimately connected. The character of military-industrial complexes, and the ties between those in advanced countries, and those in lesser nations, have clearly affected policy alignments. The range of technical, cultural, and other uncertainty syndromes that have affected Western (and Soviet) procurement patterns promise also to affect those of new local power claimants, and these in turn promise to reverberate back on the former.

The decade of the 1970s had produced total arms expenditures of $5 million million, that is, $5,000,000,000,000 (in then-current 1980 prices).[29] Yet, as one commentator was moved to note: "What makes the arms race a global folly is that all countries are now buying greater and greater insecurity at higher and higher cost."[30]

The mobility (as suggested by MX variants) and small size (exemplified by cruise missiles) of the strategic systems now being deployed by Washington, and apparently emulated by Moscow, meant that arms control aspirations of the future would have to command extraordinary political will and power. Technical verification problems were being compounded at an alarming rate. It was a trend that could give succor only to opponents of arms control. The direction of superpower research endeavors was scarcely less disturbing. Research and development of the holy grail of effective missile defenses promised, even if unsuccessful, to create dangerous uncertainties and temptations to preempt.[31] The trend toward tactical utilization of strategic weaponry (exemplified by the Schlesinger Doctrine and by PD 59) and the development of nuclear warhead variants supposedly free from the collateral damage implications of traditional technologies (such as the "neutron bomb") were equally disturbing to many. Both implicitly lowered the threshold of nuclear weapons utilization, and hence suggested increased likelihood that the escalatory spiral would be unleashed. So also with theoretical accuracy claims that ignored the impossibility of fully compensating for the gravitational and atmospheric discrepancies between wartime and peacetime trajectory environments; they encouraged first-strike specters, which inevitably led to consideration of "launch-on-warning" strategies and the attendant risk, minimal but ever-present, of an accidental Armageddon caused by warning system malfunctions. At the same time, nuclear proliferation projections promised to resurrect the dangerous "temptation syndrome" of the late 1950s and early 1960s; new nuclear arsenals would be primitive, vulnerable, and unable to remain on permanent alert (see Chapter 2)—they would have to be used first, in an initiating strike, lest they be destroyed.

The arms dynamic could obviously not be exorcised in the real world of the 1980s. There were too many "legitimate" uncertainties. But if far more substantive efforts toward at least limited control did not soon take hold, the odds on averting catastrophe would not look good to either the professional gambler, the mathematician, or the historian. Apathy and resignation were attitudes the world could no longer afford. In January 1981 *The Bulletin of the Atomic Scientists* moved its masthead "doomsday clock" up to just four minutes before midnight.

NOTES

1. Ruth Leger Sivard, *World Military and Social Expenditures 1980*, WMSE Publications, Leesburg, Va., 1980.
2. Alva Myrdal, *The Game of Disarmament—How the United States and Russia Run the Arms Race*, Spokesman, London, 1980, P.13.
3. See Chapter 1; and *A Time to Disarm*, United Nations Association in Canada, 1978.
4. "The revolving door through which retired Admirals and Generals slide from the Pentagon to defense contractors is so wide and well-oiled that, from 1970 to 1979, 1,455 former military employees were hired by a mere eight companies. . . . The expertise brought by these individuals is not only technical but political—information on and access to policy-making that helps create a closed network in a community of shared assumptions." See *The Iron Triangle*, report by the Council on Economic Priorities, New York, 1981, and Colman McCarthy's syndicated column "An Ex-Navy Admiral Fights to Stop Nuclear Arms Race," *Washington Post*, reprinted in *Miami Herald* 20 October 1981. (Laws against conflict of interest are rarely enforced, and when they are, the penalty is derisory: "Some people have their retirement pay reduced. . . . But that's a rare case.")
5. "Dwight D. Eisenhower, 1960," *Public Papers of the Presidents*, p. 1038, National Archives and Records Service, Government Printing Office, Washington, D.C., 1961.
6. Kosta Tsipis, "Scientists and Weapons Procurement," *The Bulletin of the Atomic Scientists*, June 1980, p. 42. The author goes on to explain why he considers the MX system "a veritable Rube Goldberg—if it were not deadly serious it would be downright silly." Scathing criticism is then directed at army and navy actively charged particle beam weapons programs, "which as any physics undergraduate with a course in electromagnetic theory can show are unworkable even in principle, to say nothing of their chances of working as weapons in a hostile environment of countermeasures." See also Chapter 3.
7. See, e.g., C. G. Jacobsen, *Soviet Strategic Interests and Canada's Northern Sovereignty*, Canadian Department of National Defense ORAE E-M Paper, No. 4, 1978.
8. Stockholm International Peace Research Institute, *SIPRI Yearbook 1975*, Almqvist & Wiksell, Stockholm, 1975, p. 100.
9. Sivard, *World Military and Social Expenditures.*
10. Frank Barnaby, "Arms Industry—a Seller's Market," *The Bulletin of the Atomic Scientists*, May 1981, p. 11; see also Anthony Sampson, *The Arms Bazaar*, Hodder and Stoughton, London, 1978.
11. The 1975 figure from Myrdal, *The Game of Disarmament,* p. 155; the 1980 figure is calculated on the basis of numbers provided by *Soviet Military Power,* US Department of Defense, Washington, D.C., 1981, p. 73. It should be noted that the Soviet total provided by the latter publication includes specialists excluded from the American total; the Soviet number lumps Ph.D.-level professionals with graduates of more minimalist programs (in medical terms, one might say it lumps surgeons with "barefoot doctors," and some midwives. . .). The "Annual Report from the Stockholm International Peace Research Institute," published in *The Bulletin of the Atomic Scientists*, September 1979, noted that "more than one-half of the world's physical and engineering scientists, for example, work *only* for the military" (emphasis added). The conservative estimate of 50 percent has been retained in the text. But the quantum leap in "military" scientist totals between 1975 and 1980 invites the inference that the actual percentage is today considerably higher.
12. C. T. Phan, ed., *World Politics 81/82*, Dushkin Publishing, Guilford, Conn., 1981, p. 66.

13. Ruth Leger Sivard, "World Military Expenditures 1980," in Phan, *World Politics 80/82,* p. 63.
14. European Economic Community data, cited in R. Brickman, "The Price of Success: Science and Technology in Europe," *The Bulletin of the Atomic Scientists,* September 1980, p. 31.
15. Sivard, in *World Military and Social Expenditures 1980*, provides a table of comparison, "Military Burden and National Productivity 1960–1973."
16. See, e.g., "The Brain Drain," in Alva Myrdal's *The Game of Disarmament,* pp. 155–156. And note John Kenneth Galbraith, "The Economics of the Arms Race—and After," *The Bulletin of the Atomic Scientists*, June/July 1981, pp. 13–16.
17. Quoted by Myrdal in *The Game of Disarmament*, p. 158.
18. Frank Barnaby, "Military-Scientists," *The Bulletin of the Atomic Scientists*, June/July 1981, p. 11; see also F. A. Long and J. Reppy, *The Genesis of New Weapons: Decision-Making for Military R&D*, Pergamon Policy Studies, Pergamon Press, Elmsford, N.Y., 1980.
19. C. G. Jacobsen, *Soviet Strategy–Soviet Foreign Policy*, Glasgow University Press, Robert MacLehose, 2nd ed., 1974 (US agents, Humanities Press, Atlantic Highlands, N.J.); see especially Chapter 8, "Military-Civilian Integration in the USSR," pp. 170–199.
20. See, e.g., transcript of *World Special: The Red Army*, PBS (Public Broadcasting System), Boston, 1981.
21. Tsipis, "Scientists and Weapons Procurement," p. 42.
22. See, e.g., W. Moskoff, "CIA Publications on the Soviet Economy," *Slavic Review*, Summer 1981, pp. 269–272; F. D. Holzman, "Are the Soviets Really Outspending the US on Defense?" *International Security*, Spring 1980; and J. O. Grady, "The CIA and the Soviet Military Budget . . . ," unpublished CDA analysis, Summer 1981.
23. According to Kosta Tsipis in "Scientists and Weapons Procurement": "The key reason, both for the persistently poor technical judgement evidenced in consecutive generations of ineffectual, unreliable or unworkable weapons systems and the absence of negative public or Congressional response to it appears to be the dearth of independent academic scientists and technical experts interested in defense matters. . . . Academic scientists are either indifferent or have drifted away from military technical matters, alienated by the Vietnam misadventure, extirpated from the community of scientific advisors to the Administration during the Nixon years and have not returned. . . ."
24. "Net Import Reliance as a Percent of Apparent Consumption," charted in H. Brown, Secretary of Defense, *Department of Defense Annual Report Fiscal Year 1982*, Government Printing Office, Washington, D.C., January 1981, p. 22.
25. D. E. Fink, "Availability of Strategic Materials Debated," Technical Survey, *Aviation Week and Space Technology*, 5 May 1980, pp. 42–46.
26. Brown, "Net Import Reliance," p. 22.
27. *Manchester Guardian Weekly,* 13 September 1981, p. 1; see also, e.g., P. Claude, "History Is With Us," *Le Monde*, reprinted in *Manchester Guardian Weekly*, 23 August 1981, p. 12. America's "neutrality" toward South Africa's September 1981 invasion of southern Angola ("It is not our task to choose between black and white in South Africa. The Reagan Administration has no intention of destabilizing South Africa to curry favour elsewhere") contrasted with the British government's statement that the operation was "highly dangerous for stability in southern Africa." Former Conservative British Prime Minister Edward Heath elaborated: "What makes South Africa so unique in the modern world is that the debasement of human rights has become institutionalised, enshrined in law and even sanctified by reli-

gious doctrine. No Western country with a history of colonialism or with a multi-racial society could ever support such a system of legislative discrimination.

"To do so would not only violate our most deeply held principles—it would also have unimaginable consequences for racial harmony at home. It would turn allies and friends throughout the world against the West. It would bitterly divide the alliance at a time when unity has never been more important. And it would portray the Soviet Union as the friend of the oppressed in Southern Africa and the West as their enemy.

"The result would be to facilitate, even to legitimise Soviet interference in Africa and in other conflicts or regions around the world in which the West is engaged.

"Unless and until the dismantlement of apartheid is assured, it would be a grave mistake for South Africa to base her strategy on the assumption that when the chips are down the West will stand with her." For transcipts, see *Manchester Guardian Weekly*, 6 September 1981.

28. See William Epstein, "Ban on the Production of Fissionable Material for Weapons," *Scientific American*, July 1980, for analysis of 40 presently non-nuclear states believed capable of procuring nuclear arsenals within 10 years.
29. SIPRI, World Armaments and Disarmament: *SIPRI Yearbook 1981*, Taylor & Francis, London, and Oelgeschlager, Gunn & Hain, Cambridge, Mass., 1981, p. 4.
30. Myrdal, *The Game of Disarmament*, p. 7.
31. R. Garwin, "Are We on the Verge of a New Arms Race in Space?" Special Report, *The Bulletin of the Atomic Scientists*, May 1981.

Addendum

The Effects of (Just) One Nuclear Bomb

Michael Shuman, winner of *The Bulletin of the Atomic Scientists'* Rabinowitch Prize in 1980, notes:

> A single one-megaton airburst could pulverize a 50 square mile area (about the size of Washington, D.C.), incinerate 100 square miles, inflict second-degree burns over 250 square miles and disperse assuredly lethal doses of fallout over 600 to 1000 square miles. An additional 4000 square miles would receive significant amounts of radioactive contamination, causing enormous numbers of cancers, birth defects and other radiation diseases—crowding hospitals, asylums, soup lines, prisons and cemeteries for decades to come. The Hiroshima blast, which damaged only 6 percent(!) of the area which a one-megaton blast would affect, killed over 100,000 people and traumatized the survivors and their offspring permanently. A single one-megaton blast is nearly all the firepower expended in World War II.[1]

H. Jack Gieger, much-honored Professor at the School of Biomedical Education of City College of New York, elaborates:

> At one megaton—a small weapon by contemporary standards—we are trying to imagine 70 simultaneous Hiroshima explosions. At 20 mega-

tons we are trying to imagine 1,400 Hiroshima bombs detonated at the same moment in the same place. . . . At the time of Hiroshima . . . the world's total arsenal comprised two or three weapons; today . . . the total arsenal is—conservatively—in excess of 50,000 warheads.

. . . close to *10 million* people [would be] killed or seriously injured, for example, *in consequence of a single 20 megaton explosion* and the resulting firestorm on the New York metropolitan area. . . . In San Francisco. . . the [US Arms Control and Disarmament Agency] calculates that a single one-megaton air burst would kill 624,000 persons and seriously injure and incapacitate 306,000. A single 20-megaton air burst would kill 1,538,000 and seriously injure 738,000.

These figures are *serious understatements*, however. They are based on a census population distribution, that is, they make the implicit assumption that everyone is at home, when in fact a population-targetted attack is much likelier to occur on a weekday during working hours, when the population is concentrated in central-city areas closest to ground zero. And they do not allow for the probability of a firestorm or mass conflagration as the secondary consequence of a nuclear attack. A firestorm. . .may burn for days, with ambient temperatures exceeding 800° centigrade. It increases the lethal area *five-fold.*

[After a *single* San Francisco explosion] burn casualties would exceed by a factor of 10 or 20 the capacity of *all the burn-care centres in the United States*. . . . Most will die without even the simple administration of drugs for the relief of pain.[2]

But the consequences of a single explosion may be even worse. In Europe many military facilities are located near civilian nuclear reactors (and there are similar locales in the United States and the USSR). A bomb detonated on a reactor, whether by state or terrorist, by design or accident, would have even more cataclysmic consequences. Kosta Tsipis, a professor at MIT and senior associate of the Stockholm International Peace Research Institute, presented the relevant graphs in a 1981 *Scientific American* article: "Attack on a single reactor with a single [one-megaton] nuclear weapon could devastate a substantial part of Europe." A month after an attack on a Stuttgart reactor, in South Germany, the *still uninhabitable* area would cover a third of West Germany, including its industrial heartland, and it "might extend well into the UK" (depending on prevailing winds):

An area of 180 square miles would continue for *more than a century* to expose any occupant to a dose of at least 2 rem per year [This dose is more than 10 times the maximum recommended by the US Environmental

> Protection Agency]. Such an area would be a permanent monument to the catastrophe.[3]

Finally, diverting one's gaze from the human calamity that would attend a bomb explosion, one must note that just one detonation might also have disastrous military consequences, over and above those associated with target impact. The phenomenon of EMP (the electromagnetic pulse) is poorly understood. But it is clear that it could wreak havoc on military plans and designs, by knocking out the sophisticated communication systems that lie at the heart of today's military-strategic concepts:

> A single shot unleashed 300 miles above the atmosphere and incapable of producing any other damage on Earth could disrupt unprotected electronic systems throughout Europe, across North America, or even across the great Russian hinterland. Emergency radio systems are highly vulnerable, as are telephone lines, unprotected computers, and *virtually all equipment containing sensitive electronics and wire circuitry.*
>
> The problems are increased by the second electromagnetic effect—the disruption of ionized layers in the atmosphere. A one-megaton explosion produces roughly the same number of electron pairs as exist in the entire atmosphere. These disrupt the upper atmospheric layers on which long-range communication—including radar signals—depend.
>
> . . . however carefully they are designed, systems intended to carry doomsday messages are a costly pretence . . . . every apparent advance in performance of electronic systems increases vulnerability to EMP.[4]

NOTES

1. Michael Shuman, "The Mouse That Roared," *The Bulletin of the Atomic Scientists*, January 1981, pp. 15–22.
2. H. Jack Geiger, M.D., "The Illusion of Survival," *The Bulletin of the Atomic Scientists*, June/July 1981, pp. 16–20.
3. Steven A. Fetter and Kosta Tsipis, "Catastrophic Releases of Radioactivity," *Scientific American*, April 1981, pp. 41–47. Analyzing the consequences of a bomb explosion and a catastrophic reactor accident separately, the authors demonstrate why the former must be viewed with the greater horror. The lethal zone created by a one-megaton blast would be 400 times larger. Furthermore, "the probability of the detonation of a nuclear weapon . . . in the next 10 years is far greater than the probability of the catastrophic meltdown of a nuclear reactor." As George Kistiakowsky, one of the "fathers" of American nuclear programs, once mused, to direct

one's concern to civil nuclear power rather than to world arsenals is like worrying about a wart on your ass when you have cancer!

4. Anthony Tucker, "The Pentagon's Achilles Heel," (London and Manchester) *Guardian*, 2 July 1981. Tucker suggests that EMP gives ironic advantage to technologically less sophisticated Soviet systems. One must stress, however, that neither side fully understands the phenomenon. It was not appreciated in early days of atmospheric testing; even if it had, the communications and radar systems of the time were too crude compared to their later cousins to allow for confident extrapolations of data relevance. The ban on atmospheric testing guarantees a high degree of mutual uncertainty.

Index

C

D

E

H

I

J

K

L

M

T

U

V

W

Y

Z

About the Author

Carl G. Jacobsen is Professor of International Studies, Director of Soviet Studies, and Coordinator of the Strategic Studies and National Security Program at the Center for Advanced International Studies, University of Miami. He is also Adjunct Professor of the Institute of Soviet and East European Studies, Carleton University, Ottawa. He received his Ph.D. from the Institute of Soviet and East European Studies in Glasgow, Scotland, in 1971, after having spent a year in Moscow as a British Council exchange scholar.

A frequent government consultant, Dr. Jacobsen is the author of numerous articles and books on Soviet foreign policy, Soviet strategy, Soviet–Chinese relations, and arms control. His most recent books are *Soviet Strategic Initiatives: Challenge and Response* (1979) and *Sino-Soviet Relations Since Mao: The Chairman's Legacy* (1981).